Precision Farming Technologies for Monitoring Livestock and Poultry

Precision Farming Technologies for Monitoring Livestock and Poultry

Guest Editors

Ramesh Bahadur Bist

Dongyi Wang

Lilong Chai

Yijie Xiong

MDPI

Basel • Beijing • Wuhan • Barcelona • Belgrade • Novi Sad • Cluj • Manchester

Guest Editors

Ramesh Bahadur Bist
Department of Biological and Agricultural Engineering
North Carolina State University
Raleigh
USA

Dongyi Wang
Department of Biological and Agricultural Engineering
University of Arkansas
Fayetteville
USA

Lilong Chai
Department of Poultry Science
University of Georgia
Athens
USA

Yijie Xiong
Department of Biological Systems Engineering
University of Nebraska-Lincoln
Lincoln
USA

Editorial Office
MDPI AG
Grosspeteranlage 5
4052 Basel, Switzerland

This is a reprint of the Special Issue, published open access by the journal *AgriEngineering* (ISSN 2624-7402), freely accessible at: https://www.mdpi.com/journal/agriengineering/special_issues/4D26VN342X.

For citation purposes, cite each article independently as indicated on the article page online and as indicated below:

Lastname, A.A.; Lastname, B.B. Article Title. *Journal Name* **Year**, *Volume Number*, Page Range.

ISBN 978-3-7258-7633-4 (Hbk)
ISBN 978-3-7258-7634-1 (PDF)
https://doi.org/10.3390/books978-3-7258-7634-1

Cover image courtesy of Ramesh Bahadur Bist

Contents

About the Editors vii

Preface ix

Ramesh B. Bist, Dongyi Wang, Lilong Chai and Yijie Xiong
Precision Farming Technologies for Monitoring Livestock and Poultry
Reprinted from: *Agriengineering* **2026**, *8*, 64,
https://doi.org/10.3390/agriengineering8020064 1

Reagan Lewis, Teun Kostermans, Jan Wilhelm Brovold, Talha Laique and Marko Ocepek
Automated Body Condition Scoring in Dairy Cows Using 2D Imaging and Deep Learning
Reprinted from: *Agriengineering* **2025**, *7*, 241,
https://doi.org/10.3390/agriengineering7070241 6

Ramesh Bahadur Bist, Andi Asnayanti, Anh Dang Trieu Do, Yang Tian, Chaitanya Pallerla, Dongyi Wang and Adnan A. K. Alrubaye
Automated Detection of Kinky Back in Broiler Chickens Using Optimized Deep Learning Techniques
Reprinted from: *Agriengineering* **2025**, *7*, 415,
https://doi.org/10.3390/agriengineering7120415 22

Xiao Yang, Ramesh Bahadur Bist, Sachin Subedi, Zihao Wu, Tianming Liu, Bidur Paneru and Lilong Chai
A Machine Vision System for Monitoring Wild Birds on Poultry Farms to Prevent Avian Influenza
Reprinted from: *Agriengineering* **2024**, *6*, 211,
https://doi.org/10.3390/agriengineering6040211 36

Edmanuel Cruz, Miguel Hidalgo-Rodriguez, Adiz Mariel Acosta-Reyes, José Carlos Rangel, Keyla Boniche and Franchesca Gonzalez-Olivardia
ACMSPT: Automated Counting and Monitoring System for Poultry Tracking
Reprinted from: *Agriengineering* **2025**, *7*, 86,
https://doi.org/10.3390/agriengineering7030086 51

Isis Mariana Dombrowsky Leal Pasian, Robson Mateus Freitas Silveira, Jessica Nacarato Reple, Hilton Tadeu Zarate Couto and Iran José Oliveira da Silva
Efficiency and Reliability of Broiler Weighing Methods in Commercial Environments: A Comparative Evaluation
Reprinted from: *Agriengineering* **2025**, *7*, 141,
https://doi.org/10.3390/agriengineering7050141 75

Glauber da Rocha Balthazar, Robson Mateus Freitas Silveira and Iran José Oliveira da Silva
Design and Prototyping of a Robotic Structure for Poultry Farming
Reprinted from: *Agriengineering* **2025**, *7*, 233,
https://doi.org/10.3390/agriengineering7070233 84

Maria Angela de Souza, Fernanda Campos de Sousa, Alex Lopes da Silva, Thauane Cordeiro Soares, Charles Paranhos Oliveira, Ricardo Brauer Vigoderis, et al.
Effect of Green Roofs on the Thermal Environment of Prototype Broiler Houses
Reprinted from: *Agriengineering* **2025**, *7*, 16,
https://doi.org/10.3390/agriengineering7010016 107

Shengyu Zhou, Tanner Thornton, Hao Gan, Tom Tabler, Mustafa Jaihuni, Xiaojuan Zhu and Yang Zhao
Impact of Stocking Density on Welfare and Performance of Ross 708 and Cobb 700 Broilers
Reprinted from: *Agriengineering* **2024**, *6*, 271,
https://doi.org/10.3390/agriengineering6040271 . . . . **130**

About the Editors

Ramesh Bahadur Bist

Ramesh Bahadur Bist is an Assistant Professor at North Carolina State University (NC State) specializing in AI, computer vision, robotics, and sensing technologies for precision livestock and poultry farming. He earned his Ph.D. in Poultry Science (2024) from the University of Georgia under Dr. Lilong Chai, an M.S. in Animal Science (2018) from Arkansas State University, and a B.S. in Animal Science (2014) from Tribhuvan University, Nepal. In 2024, he also joined the University of Arkansas as a postdoctoral fellow, where he conducted research in computer vision and robotics. His research focuses on developing data-driven tools for animal health, behavior, welfare monitoring, autonomous unmanned vehicles, and rapid non-destructive food quality detection. Dr. Bist has published 100+ scientific publications, contributed to major journals in agriculture and AI, and currently serves as a Guest Editor and active reviewer for multiple leading scientific journals.

Dongyi Wang

Dongyi Wang is the director of the Smart Agriculture and Food Engineering (SAFE) lab. SAFE lab's primary research focus is developing and applying advanced artificial intelligence, robotics, and machine vision techniques for modern agrifood manufacturing to protect the quality, safety, and welfare of crops, animals, food, and workers. SAFE also commits to developing an integrated education program to build an AI/Robotics-skilled workforce, addressing the needs in both the state and national agrifood industries. Dr. Wang has authored 40+ peer-reviewed publications, delivered 50+ presentations, held 5 U.S. patents, and is an active member in ASABE and IEEE.

Lilong Chai

Lilong Chai is an Associate Professor & Engineering Specialist in the Department of Poultry Science at the University of Georgia (UGA). He is a UGA Aspire Fellow & Steering Committee Member of the UGA Institute for Integrative Precision Agriculture. Chai's research and extension interests include animal environment, precision poultry farming, and poultry health and welfare. Chai is an author/coauthor of 300 scientific publications (including 100 journal articles, 150 conference papers/abstracts, and 60 extension articles) and a PI/Co-PI on 45 grants/contracts funded by USDA-NIFA, Egg Industry Center, Georgia Research Alliance, etc. He currently serves as the Coordinator of the Georgia Precision Poultry Farming Conference and the Georgia Layer Conference, two annual extension poultry training programs at UGA, and as Chair of the 4th US Precision Livestock Farming Conference (2025–2027). Chai's research and extension efforts have been recognized with 20 professional awards and honors, including the 2025 American Egg Board Research Award and the 2021 ASABE Sunkist Young Designer Award.

Yijie Xiong

Yijie Xiong is an Assistant Professor and Extension Precision Livestock Management Specialist in the Department of Animal Science at the University of Nebraska Lincoln, with a split appointment in Biological Systems Engineering. Her research focuses on precision livestock management, wearable sensing systems, animal behavior monitoring, and data analytics to improve animal health, welfare, and production efficiency. She develops and validates practical technologies for real-world livestock systems and leads extension programs that help producers adopt precision management tools. She earned her Ph.D. in Agricultural Engineering from the University of Illinois at Urbana-Champaign and previously served as a Guest Editor for the *Animals* Special Issue on automated monitoring of livestock and poultry with machine learning technology.

Preface

The Reprint "Precision Farming Technologies for Monitoring Livestock and Poultry" brings together research focused on the rapid changes in animal agriculture enabled by advanced monitoring and smart systems. This Reprint covers sensing technologies, data analytics, artificial intelligence, robotics, and decision-support tools that help improve livestock and poultry production. It examines both scientific progress and real-world applications, highlighting solutions that work well on farms.

This Reprint was created to address the need for improved animal welfare, greater production efficiency, environmental sustainability, and system resilience, especially amid challenges such as labor shortages, disease risks, and climate change. Precision farming technologies are now key parts of modern production, not just experimental ideas. By sharing work from animal scientists, engineers, and data scientists, this Reprint shows how working together leads to real, useful solutions.

This Reprint is for researchers, industry professionals, extension specialists, technology developers, and graduate students who want to advance precision monitoring in livestock and poultry systems. The collected studies aim to serve as a useful reference and to inspire further innovation in smart animal agriculture.

Ramesh Bahadur Bist, Dongyi Wang, Lilong Chai, and Yijie Xiong
Guest Editors

AgriEngineering

MDPI

Editorial

Precision Farming Technologies for Monitoring Livestock and Poultry

Ramesh B. Bist [1,*], Dongyi Wang [2], Lilong Chai [3] and Yijie Xiong [4]

1 Department of Biological and Agricultural Engineering, North Carolina State University, Raleigh, NC 27695, USA
2 Department of Biological and Agricultural Engineering, University of Arkansas, Fayetteville, AR 72701, USA
3 Department of Poultry Science, University of Georgia, Athens, GA 30602, USA; lchai@uga.edu
4 Department of Animal Science, University of Nebraska-Lincoln, Lincoln, NE 68583, USA
* Correspondence: rbbist@ncsu.edu

Precision farming technologies are changing how livestock and poultry are managed, progressing from occasional checks to ongoing, data-driven monitoring [1–4]. Since the mid-2010s, advances in sensors, computing, and artificial intelligence (AI) have created data-rich farm environments, enabling the more accurate, faster tracking of animal health, behavior, welfare, and productivity. This shift is now a necessity, not just a result of innovation. Farms around the world are dealing with labor shortages [5,6], more complex disease risks [7], climate-related stress [8], and higher expectations of animal welfare, food safety, and transparency [9]. Because of these challenges, precision technologies are becoming vital for building systems that can rapidly identify problems, react quickly, and improve results while also lessening environmental impact.

The current progress builds on years of research in animal biology, welfare science, engineering, and computing. Early precision livestock farming mainly focused on controlling the environment and tracking basic performance, but systems are now more focused on the animals themselves. Important signals of welfare and health are directly collected from the animals, and this information is applied for implementing useful actions [10,11]. However, real challenges face the wider adoption of newer technologies. Many monitoring tools are expensive, hard to scale, or unreliable in real farm settings. In addition, lighting, dust, animal movement, and varied housing designs can also cause issues. Predictive models often perform well in controlled tests but are not generalizable across farms, breeds, ages, and production systems. High-quality labeled data on behaviors and welfare are also lacking, especially for early or mild cases where early action is most important. These challenges emphasize the need for technologies that are not only accurate but also dependable, interpretable, and practical for real farm conditions. In addition, farmers need access to research-based, up-to-date information, yet their opportunities to connect directly with extension specialists and educators are often limited. In this context, tools such as PoultryTalk (built on a retrieval-augmented generation (RAG) framework) are essential for delivering the right poultry-specific knowledge to the right person at the right time 24/7 [12]. Expanding the availability of such domain-focused systems across the industry can strengthen decision-making and improve outcomes throughout the poultry supply chain.

This Special Issue, "Precision Farming Technologies for Monitoring Livestock and Poultry", features work that addresses these challenges and shows how the field is transitioning from early ideas to practical, production-ready solutions. A key theme is combining engineering advances with biological needs, ensuring that sensing and AI systems focus

 https://doi.org/10.3390/agriengineering8020064

on real improvements such as better welfare, disease prevention, higher performance, and more efficient management. The studies cover a wide range of topics from poultry and dairy to larger facilities and biosecurity issues. Together, these studies show that effective monitoring needs to operate at multiple levels, including tracking individual animal health and welfare, monitoring group trends, and accounting for farm-wide factors such as the environment and operations.

Several studies in this Special Issue show how computer vision and deep learning are setting new standards for automated welfare checks and early health detection (Table 1). For example, Lewis et al., 2025 (contribution 1) used 2D imaging and deep learning to automate body condition scoring in dairy cows, thereby reducing subjectivity and supporting both productivity and welfare. In poultry, Bist et al., 2025 (contribution 2) developed deep learning methods to automatically detect kinky back in broiler chickens, highlighting the need for early, scalable methods for detecting skeletal problems that are often missed in the early stages. At the farm biosecurity level, Yang et al. 2024 (contribution 3) developed a machine vision system to monitor wild birds on poultry farms, helping prevent avian influenza and demonstrating how vision-based systems can support proactive disease management. These studies show that AI can be used for more than general monitoring. AI can help detect specific welfare and disease issues, enabling earlier action and better outcomes.

Table 1. Summary of publications and key contributions in this Special Issue.

Article	Application/ Task	Models/ Methods	Sensors/ Data	Platform/ System	Key Contributions	Authors
1	Dairy cow body condition scoring	Deep-learning-based image analysis	2D RGB images	Fixed camera system	Developed a cost-effective automated and non-invasive body condition scoring framework for dairy cows using 2D imaging and deep learning	Lewis et al., 2025 (contribution 1)
2	Detection of kinky back in broiler chickens	Optimized deep learning models	RGB image data	Vision-based inspection system	Developed an automated vision-based system for detecting skeletal deformities in broiler chickens to improve welfare and management	Bist et al., 2025 (contribution 2)
3	Wild bird monitoring for avian influenza prevention	Machine vision and object detection	Camera image data	On-farm monitoring system	Introduced a machine vision system to detect and monitor wild birds on poultry farms as a biosecurity measure against avian influenza	Yang et al. 2024 (contribution 3)
4	Automated poultry counting and tracking	Computer vision algorithms	Video and image data	Automated counting and monitoring system	Designed and validated an automated cost-effective system for real-time poultry counting and movement monitoring to support precision livestock management	Cruz et al. 2025 (contribution 4)

https://doi.org/10.3390/agriengineering8020064

Table 1. *Cont.*

Article	Application/ Task	Models/ Methods	Sensors/ Data	Platform/ System	Key Contributions	Authors
5	Evaluation of broiler weighing methods	Statistical and comparative analysis	Weight measurement data	Commercial poultry houses using suspended platforms	Compared the efficiency and reliability of different broiler weighing methods under commercial production conditions	Pasian et al. 2025 (contribution 5)
6	Robotic structure design for poultry farming	Mechanical design and prototyping	Sensor-integrated components	Robotic platform	Designed and prototyped a robotic structure aimed at increasing automation and labor efficiency in poultry farming	Da Rocha Balthazar et al. 2025 (contribution 6)
7	Thermal environment assessment in broiler houses	Environmental monitoring and analysis	Temperature and environmental sensors	Prototype broiler houses	Evaluated the impact of green roofs on the thermal environment of broiler houses, highlighting potential energy and welfare benefits	De Souza et al. 2025 (contribution 7)
8	Effects of stocking density on broiler welfare	Experimental and statistical analyses	Performance and welfare indicators	Commercial broiler production systems	Assessed the effects of stocking density on welfare and performance of different broiler strains to support evidence-based management decisions	Zhou et al. 2024 (contribution 8)

This Special Issue also shows that achieving success in precision farming relies on more than just algorithms: it depends on systems, infrastructure, and adapting to the environment. Cruz et al. 2025 (contribution 4) introduced ACMSPT, an automated system using AI and edge computing for poultry tracking that addresses real challenges such as limited connectivity and the need for scalable monitoring. Pasian et al. 2025 (contribution 5) examined how well broiler weighing methods perform in commercial settings, stressing that reliable growth monitoring is key to management and that automated systems must prove themselves under real-world conditions. Da Rocha Balthazar et al. 2025 (contribution 6) designed and tested a robotic structure for poultry farming, showing how robotics can extend precision monitoring and automation. De Souza et al. 2025 (contribution 7) studied green roofs in broiler house prototypes, showing that facility design can affect thermal comfort and environmental stability, which are important for welfare, performance, and sensor reliability. Zhou et al. 2024 (contribution 8) studied how stocking density affects welfare and performance in broilers, reminding us that precision farming must be based on biological understanding and management science to truly improve welfare and productivity, not only collect data.

Looking ahead, the studies in this Special Issue highlight the next steps for precision livestock and poultry farming by making systems more reliable on farms, using multiple data types, supporting real-time decision-making, making AI easier to understand, and restraining costs. Future research should focus on testing models across different farms and adapting them to new situations, so the models will work well with various housing, breeds, and management styles. Combining data from vision, sound, temperature, environment,

https://doi.org/10.3390/agriengineering8020064

and possibly wearable sensors should advance from theory to practical systems that work reliably under real-world conditions. Real-time decision support should link detection to action with alerts, advice, and automated responses that fit into producers' routines. Ensuring AI is understandable will help build trust and allow for better biological insights. Widespread adoption will also depend on practical engineering, such as the development of affordable sensors, edge computing, energy-efficient models, and systems that are easy to maintain. In summary, this Special Issue demonstrates that precision farming technologies are becoming essential in animal agriculture and underscores the need for collaboration among animal scientists, engineers, data scientists, and industry partners. We thank all the authors and reviewers for their contributions and hope this volume serves as both a reference and a catalyst for accelerating the development of reliable, scalable, and effective solutions for monitoring livestock and poultry worldwide.

Conflicts of Interest: The authors declare no conflicts of interest.

List of Contributions:

1. Lewis, R.; Kostermans, T.; Brovold, J.W.; Laique, T.; Ocepek, M. Automated Body Condition Scoring in Dairy Cows Using 2D Imaging and Deep Learning. *AgriEngineering* **2025**, *7*, 241.
2. Bist, R.B.; Asnayanti, A.; Do, A.D.T.; Tian, Y.; Pallerla, C.; Wang, D.; Alrubaye, A.A.K. Automated Detection of Kinky Back in Broiler Chickens Using Optimized Deep Learning Techniques. *AgriEngineering* **2025**, *7*, 415. https://doi.org/10.3390/agriengineering7120415.
3. Yang, X.; Bist, R.B.; Subedi, S.; Wu, Z.; Liu, T.; Paneru, B.; Chai, L. A Machine Vision System for Monitoring Wild Birds on Poultry Farms to Prevent Avian Influenza. *AgriEngineering* **2024**, *6*, 3704–3718. https://doi.org/10.3390/agriengineering6040211.
4. Cruz, E.; Hidalgo-Rodriguez, M.; Acosta-Reyes, A.M.; Rangel, J.C.; Boniche, K.; Gonzalez-Olivardia, F. ACMSPT: Automated Counting and Monitoring System for Poultry Tracking. *AgriEngineering* **2025**, *7*, 86.
5. Pasian, I.M.D.L.; Silveira, R.M.F.; Reple, J.N.; Couto, H.T.Z.; da Silva, I.J.O. Efficiency and Reliability of Broiler Weighing Methods in Commercial Environments: A Comparative Evaluation. *AgriEngineering* **2025**, *7*, 141. https://doi.org/10.3390/agriengineering7050141.
6. da Rocha Balthazar, G.; Silveira, R.M.F.; da Silva, I.J.O. Design and Prototyping of a Robotic Structure for Poultry Farming. *AgriEngineering* **2025**, *7*, 233.
7. de Souza, M.A.; de Sousa, F.C.; da Silva, A.L.; Soares, T.C.; Oliveira, C.P.; Vigoderis, R.B.; da Baêta, F.C.; de Tinôco, I.F.F. Effect of Green Roofs on the Thermal Environment of Prototype Broiler Houses. *AgriEngineering* **2025**, *7*, 16.
8. Zhou, S.; Thornton, T.; Gan, H.; Tabler, T.; Jaihuni, M.; Zhu, X.; Zhao, Y. Impact of Stocking Density on Welfare and Performance of Ross 708 and Cobb 700 Broilers. *AgriEngineering* **2024**, *6*, 4739–4751. https://doi.org/10.3390/agriengineering6040271.

References

1. Essien, D.; Neethirajan, S. Multimodal AI Systems for Enhanced Laying Hen Welfare Assessment and Productivity Optimization. *Smart Agric. Technol.* **2025**, *12*, 101564. [CrossRef]
2. Neethirajan, S. Digital Twins for Cows and Chickens: From Hype Cycles to Hard Evidence in Precision Livestock Farming. *Agriculture* **2026**, *16*, 166. [CrossRef]
3. Astill, J.; Dara, R.A.; Fraser, E.D.; Roberts, B.; Sharif, S. Smart Poultry Management: Smart Sensors, Big Data, and the Internet of Things. *Comput. Electron. Agric.* **2020**, *170*, 105291. [CrossRef]
4. Losacco, C.; Pugliese, G.; Forte, L.; Tufarelli, V.; Maggiolino, A.; De Palo, P. Digital Transition as a Driver for Sustainable Tailor-Made Farm Management: An Up-to-Date Overview on Precision Livestock Farming. *Agriculture* **2025**, *15*, 1383.
5. Taylor, J.E.; Charlton, D. *The Farm Labor Problem: A Global Perspective*; Academic Press: Cambridge, MA, USA, 2018; ISBN 0-12-817268-1.
6. Christiaensen, L.; Rutledge, Z.; Taylor, J.E. The Future of Work in Agriculture: Some Reflections. *World Bank Policy Res. Work. Pap.* 2020. Available online: https://ssrn.com/abstract=3560626 (accessed on 29 January 2026).

7. Huntington, B.; Bernardo, T.M.; Bondad-Reantaso, M.; Bruce, M.; Devleesschauwer, B.; Gilbert, W.; Grace, D.; Havelaar, A.; Herrero, M.; Marsh, T.L. Global Burden of Animal Diseases: A Novel Approach to Understanding and Managing Disease in Livestock and Aquaculture. *Rev. Sci. Tech. l'OIE* **2021**, *40*, 567–584.
8. Berry, H.L.; Hogan, A.; Owen, J.; Rickwood, D.; Fragar, L. Climate Change and Farmers' Mental Health: Risks and Responses. *Asia Pac. J. Public Health* **2011**, *23*, 119S–132S. [PubMed]
9. Alonso, M.E.; González-Montaña, J.R.; Lomillos, J.M. Consumers' Concerns and Perceptions of Farm Animal Welfare. *Animals* **2020**, *10*, 385. [CrossRef] [PubMed]
10. Buller, H.; Blokhuis, H.; Lokhorst, K.; Silberberg, M.; Veissier, I. Animal Welfare Management in a Digital World. *Animals* **2020**, *10*, 1779. [CrossRef] [PubMed]
11. Thomann, B.; Würbel, H.; Kuntzer, T.; Umstätter, C.; Wechsler, B.; Meylan, M.; Schüpbach-Regula, G. Development of a Data-Driven Method for Assessing Health and Welfare in the Most Common Livestock Species in Switzerland: The Smart Animal Health Project. *Front. Vet. Sci.* **2023**, *10*, 1125806. [CrossRef] [PubMed]
12. Khanal, K.; Khatiwada, B.; Afrifa, S.; Sapkota, R.; Shah, S.; Bai, F.; Bist, R.B. PoultryTalk: A Multi-Modal Retrieval-Augmented Generation (RAG) System for Intelligent Poultry Management and Decision Support. *arXiv* **2025**, arXiv:2512.08995.

AgriEngineering

MDPI

Article

Automated Body Condition Scoring in Dairy Cows Using 2D Imaging and Deep Learning

Reagan Lewis [1,2,3,*], Teun Kostermans [1,4], Jan Wilhelm Brovold [1], Talha Laique [1] and Marko Ocepek [1]

1 Department of Animal and Aquacultural Sciences, Norwegian University of Life Sciences, P.O. Box 5003, NO-1432 As, Norway; teun.kostermans@gmail.com (T.K.); jan.wilhelm.brovold@nmbu.no (J.W.B.); talha.laique@nmbu.no (T.L.)
2 Department of Animal Sciences, Egerton University, P.O. Box 536-20115, Njoro 20115, Kenya
3 Department of Animal Production, University of Juba, Juba P.O. Box 82, Central Equatoria, South Sudan
4 Department of Animal Sciences, Wageningen University, P.O. Box 338, 6700 AH Wageningen, The Netherlands
* Correspondence: reaganlouis163@yahoo.com

Abstract: Accurate body condition score (BCS) monitoring in dairy cows is essential for optimizing health, productivity, and welfare. Traditional manual scoring methods are labor-intensive and subjective, driving interest in automated imaging-based systems. This study evaluated the effectiveness of 2D imaging and deep learning for BCS classification using three camera perspectives—front, back, and top-down—to identify the most reliable viewpoint. The research involved 56 Norwegian Red milking cows at the Center for Livestock Experiments (SHF) of Norges Miljo-og Biovitenskaplige Universitet (NMBU) in Norway. Images were classified into BCS categories of 2.5, 3.0, and 3.5 using a YOLOv8 model. The back view achieved the highest classification precision (mAP@0.5 = 0.439), confirming that key morphological features for BCS assessment are best captured from this angle. Challenges included misclassification due to overlapping features, especially in Class 2.5 and background data. The study recommends improvements in algorithmic feature extraction, dataset expansion, and multi-view integration to enhance accuracy. Integration with precision farming tools enables continuous monitoring and early detection of health issues. This research highlights the potential of 2D imaging as a cost-effective alternative to 3D systems, particularly for small and medium-sized farms, supporting more effective herd management and improved animal welfare.

Keywords: precision livestock farming; cow welfare monitoring; camera perspectives; 3D systems

1. Introduction

Body condition scoring (BCS) is a crucial tool in dairy cow management, providing valuable insights into an animal's nutritional status, health, and productivity [1]. BCS is widely recognized as an essential welfare indicator, influencing reproductive efficiency, milk yield, and disease susceptibility [2]. Traditional BCS assessments rely on manual scoring systems, such as the Welfare Quality® protocol, which are subjective and prone to inconsistency among evaluators [3]. These limitations have driven the development of automated imaging technologies, which enable more objective, reliable, and continuous assessments of dairy cow body condition [4].

Advanced BCS monitoring systems utilizing 3D imaging have demonstrated high accuracy and precision by capturing depth-related body shape features in dairy cows [5]. These systems provide detailed morphological data, allowing for consistent evaluation of

 https://doi.org/10.3390/agriengineering7070241

body reserves and early detection of metabolic imbalances [6]. However, the widespread adoption of 3D imaging is limited due to high costs, complex installations, and maintenance requirements, making it less accessible to small and medium-sized farms [7]. As a result, alternative 2D imaging solutions have emerged, offering a more cost-effective and scalable approach to BCS automation.

Recent advancements in machine learning and computer vision have greatly improved the accuracy of 2D imaging-based BCS systems [7]. Although these systems do not capture depth information like 3D cameras, they utilize high-resolution images and deep learning algorithms to identify key morphological patterns linked to body condition [8]. Two-dimensional cameras are easier to integrate into existing farm infrastructure, cost less to implement, and can provide real-time, 24/7 monitoring [9]. Additionally, they can serve multiple functions, such as behavioral analysis and early detection of health issues, helping to improve farm management efficiency [10].

A significant challenge for smaller dairy farms is their limited access to costly robotic milking systems, which often feature integrated 3D imaging solutions [11]. For farms that do not use automated milking robots, standalone 2D imaging systems could be a practical alternative. These systems can be placed in key locations within barns, such as feeding stations, weighing platforms, or milking parlors, enabling farmers to continuously monitor BCS without manual assessment [11]. The potential for real-time decision-making based on automated BCS evaluations could help optimize nutritional strategies, reproductive planning, and health interventions, making dairy farming more efficient and sustainable [12].

This study aims to assess the potential of 2D imaging for BCS classification by identifying the most effective imaging angles through automated image analysis. The hypothesis is that 2D imaging, combined with deep learning methods, can offer an accurate, cost-efficient, and scalable alternative to manual BCS evaluation and costly 3D systems. This research concentrates on analyzing three camera views (front, rear, and top-down) to evaluate their effectiveness in classifying BCS. It also investigates how image preprocessing, feature extraction, and dataset augmentation techniques influence classification performance.

Recent research indicates that deep learning-based imaging can facilitate cost-effective automation of BCS scoring; however, there is still limited agreement on the best camera positioning or system design for real-time farm use. This study aims to fill these gaps. Its goal is to assess the potential of 2D imaging combined with deep learning for automatically classifying body condition scores (BCSs). Specifically, the research examines how three camera angles—front, back, and top-down—affect classification accuracy using YOLOv8. The objective is to determine the most effective imaging perspective for scalable, affordable deployment and to evaluate how image preprocessing and augmentation methods impact model performance.

2. Materials and Methods

2.1. Data Collection

2.1.1. Study Location and Duration

The study was conducted at the Animal Research Centre of the Norwegian University of Life Sciences (NMBU) in As, Norway, in accordance with legal requirements for the keeping of dairy cattle (Ministry of Agriculture and Food, 2007) [13]. Since standard cattle rearing practices were followed and no potentially harmful research was performed, no approval was required to obtain separate permission for research involving animals (Ministry of Agriculture and Food, 2015) [14]. The research took place over a two-week period from 5 November to 19 November 2024, at a dairy cattle farm equipped with an Automated Milking System (AMS). The study involved 56 Norwegian Red breed milking cows.

2.1.2. Subjects and Data Acquisition

- Cows' Access to AMS: Cows voluntarily accessed the AMS between one and four times per day, where various physiological and behavioral parameters were recorded.
- BCS Reference Standard: A 3D camera system installed above the AMS served as the reference standard for BCS evaluation.
- Additional Data Recorded: The AMS automatically logs cow identification (ID), milking session timestamps, and body weights, all organized into a structured Excel database for further analysis.

2.2. Digital Tools and Camera Setup

2.2.1. Camera Specifications

- Two-Dimensional Cameras: Three Foscam G4EP PoE 4MP cameras, each equipped with 128 GB SD cards for motion-triggered image capture.
- Three-Dimensional Camera: Installed above the AMS, capturing depth-related morphological features of cows during milking.
- Data Storage and Processing: Cameras were connected to an Ethernet network, and recording schedules were configured via the Foscam mobile application.

The camera system was installed in the AMS lane of the dairy barn, with three fixed RGB cameras strategically positioned to capture top, back, and front views of each cow as it moved through the milking stall. As shown in Figure 1, the top-view camera was mounted directly above the stall's central lane on the overhead metal frame, providing a downward angle focused on the cow's spine and loin area. The back-view camera was set up behind the cow's exit path. The front-view camera was placed near the stall entry gate, facing the cow's head and chest to capture frontal features important for body condition scoring, such as the brisket and shoulders. All cameras were secured using mounting brackets and aligned to reduce occlusion while maintaining a stable view during image collection.

Figure 1. Overview of the barn and camera placement [15].

Recent studies have identified 3D imaging systems as the gold standard for body condition scoring and morphological assessment in livestock because of their ability to accurately capture depth and contour information [16]. However, these systems are expensive and complicated to implement on commercial farms, which has encouraged research into 2D-based alternatives.

2.2.2. Camera Placement

To capture comprehensive BCS data, both 3D and 2D cameras were strategically positioned within the AMS and the weighing station after milking.

- 3D Camera Placement
    - Mounted directly above the AMS milking unit to provide a top-down view of cows for depth-based morphological analysis.
    - This 3D-based BCS evaluation served as the gold standard for validation.
- 2D Camera Placement
    - Front View Camera—Positioned at the entrance of the weighing station, capturing the head and shoulder regions.
    - Rear View Camera—Mounted behind the weighing scale, providing a backward perspective of the cow's hindquarters.
    - Top-Down Camera—Mounted above the weighing platform, providing a bird's-eye view of the cow's topline and body structure.

All captured frames from 2D cameras were systematically linked to cow ID and BCS data, ensuring a high-quality dataset for training and validating the YOLOv8-based BCS classification model.

2.3. Image Data Processing and Annotation

Image Collection and Preprocessing

1. Data Collection
    - Video footage and snapshots were collected daily from 8:00 a.m. to 8:00 p.m., with images taken whenever movement was detected.
2. Frame Extraction
    - At the end of each milking session, a screenshot was taken of each cow standing on the weighing scale to ensure a direct link to BCS and ID records.
3. Image Sorting
    - Screenshots were organized by cow ID and BCS score.
4. BCS Score Adjustment
    - Scores were rounded to the nearest half or whole grade (e.g., BCS 1.7 $\rightarrow$ 1.5; BCS 2.8 $\rightarrow$ 3.0).

Dataset Annotation and Preparation

1. Manual Annotation
    - The dataset was uploaded to Roboflow using their SAM annotation tool. In some cases, the manual polygon tool was employed to assign object detection labels to front, rear, and top views.
2. Dataset Splitting
    - After annotation, the dataset was divided into three sets
        - 70% Training Data—Used to train the YOLOv8 model.
        - 20% Validation Data—Used for hyperparameter tuning and overfitting prevention.

- 10% Testing Data—Used for final model evaluation.

The initial dataset included images across five BCS categories: 1.5, 2.0, 2.5, 3.0, and 3.5. However, classes 1.5 and 2.0 were severely underrepresented, with fewer than 30 usable annotated images combined due to the low prevalence of cows with these BCS values during the data collection period. To avoid training bias and class imbalance issues during model development, these minority classes were excluded from the final classification task. This decision was made to ensure model stability and to prevent overfitting to rare classes with insufficient feature diversity. As a result, the final model was trained on three balanced BCS classes (2.5, 3.0, 3.5), for which we ensured adequate image representation across all views.

2.4. *Object Detection Model*

2.4.1. Model Architecture and Justification

For automated BCS classification, we used YOLOv8, a cutting-edge object detection model developed by Ultralytics. YOLOv8 has three main components.

- Backbone: Extracts hierarchical image features using convolutional layers, C2f blocks, and spatial pyramid pooling.
- Neck: Fuses multi-scale feature maps, optimizing object detection across different perspectives (rear, front, and top).
- Head: Generates final predictions, including bounding-box coordinates, class labels, and confidence scores.

YOLOv8 was selected for its high detection accuracy, real-time inference capabilities, and efficient feature extraction, making it well suited for multi-view BCS assessment in dairy cows.

2.4.2. Data Preprocessing and Augmentation

A total of 1764 annotated images were collected for model development. These images represented three body condition score (BCS) classes: 521 images labeled as BCS 2.5, 648 as BCS 3.0, and 595 as BCS 3.5. Regarding imaging perspectives, the dataset included 588 back-view, 593 front-view, and 583 top-view images. A stratified split divided the dataset into training (1234 images), validation (353 images), and testing (177 images) sets, maintaining class balance across views and BCS classes.

To improve model generalization, the following image augmentation techniques were applied:

- Auto-orientation (standardized image alignment).
- Resize (all images resized to 640 × 640 pixels).
- Contrast and Brightness Normalization (to enhance feature visibility).
- Data Augmentation
    - Cropping: 0–15% random zoom.
    - Hue and Saturation Adjustments: Random variations (−6 to +6).
    - Brightness: Randomly adjusted (−15% to +15%).
    - Blur and Noise: Minor Gaussian blur and noise applied.

To address class imbalances, underrepresented BCS categories (BCS 1.5 and 2.0) were removed, leaving three final classes.

- BCS 2.5 → Class 25
- BCS 3.0 → Class 30
- BCS 3.5 → Class 35

The image annotation was performed using Roboflow, a web-based platform for annotation and dataset management, which allowed manual labeling of bounding boxes for each image. The labeled dataset was exported in a format compatible with YOLOv8. Image preprocessing and augmentation were performed with the Albumentations library (version 1.3.1), offering a comprehensive set of computer vision transformations. Preprocessing included resizing all images to 640 $\times$ 640 pixels, normalizing pixel values based on ImageNet mean and standard deviation, and maintaining consistent aspect ratios. During training, augmentation techniques such as horizontal flipping, random rotations up to $\pm 15^\circ$, brightness and contrast adjustments, CLAHE (Contrast Limited Adaptive Histogram Equalization), and random cropping were applied on the fly using the Ultralytics YOLOv8 training script (version 8.0.104), which is fully integrated with a PyTorch-based (PyTorch version 1.13.1) training environment.

YOLOv8 was chosen for this study because of its balance of speed, accuracy, and deployability. It is a state-of-the-art one-stage object detection architecture that improves on YOLOv5 by using a decoupled head, anchor-free design, and enhanced feature extraction with only a slight increase in model size. These features make it especially suitable for deployment in real-time livestock monitoring systems where latency and resource constraints are critical.

While heavier models like Faster R-CNN or RetinaNet can achieve higher accuracy on controlled benchmarks, they usually require more memory, have slower inference times, and are less practical for edge-based systems. Conversely, ultra-light models such as YOLO-Nano or MobileNet-SSD provide faster speeds but sacrifice some detection accuracy—especially in complex farm environments with overlapping features and lighting variation.

In this study, the priority was to identify a model that could provide adequate classification performance (mAP > 0.40) while remaining computationally feasible for eventual on-site or mobile deployment. YOLOv8 achieves this balance effectively, which is why it was selected. However, future work will explore benchmarking against lighter models (e.g., YOLOv5n, YOLOv7-tiny, MobileNet-SSD) and heavier ones (e.g., Efficient Det, Faster R-CNN) to quantify trade-offs in BCS classification tasks further.

2.4.3. Model Training and Optimization

Training was conducted on Google Colab, utilizing an NVIDIA Tesla T4 GPU (16 GB RAM).

- Pretrained Weights: Transfer learning was applied using MSCOCO weights.
- Optimizer: Adam optimizer with learning rate = 0.001.
- Training Setup
    - Epochs: 50
    - Batch size: 16
    - Loss functions:
        - Bounding box (BBox) loss.
        - Classification loss.
        - Distribution Focal Loss (DFL).

To assess model stability and mitigate the impact of random initialization, we conducted three independent training runs of the YOLOv8 model using identical hyperparameters (50 epochs, batch size 16, and Adam optimizer) and randomized weight initialization. Each run used the same data splits (70% training, 20% validation, 10% testing). The results were consistent across runs, with standard deviation in mAP@0.5 values <0.015 across all views and classes. While formal k-fold cross-validation was not conducted due to computational limitations, the consistent performance across trials supports the robustness of our findings.

2.4.4. Model Evaluation

The trained model's performance was evaluated using the following:

- Mean Average Precision (mAP@0.5 and mAP@0.5–0.95).
- Precision/Recall Curves.
- Confusion Matrices (misclassification analysis).
- Confidence Threshold Optimization (for front, rear, and top views).

After training, the model was deployed and tested on an Intel Core i7-12700H (32 GB RAM) for real-time performance validation. The training and validation loss plots of the models can be seen in Figures 2–4 below.

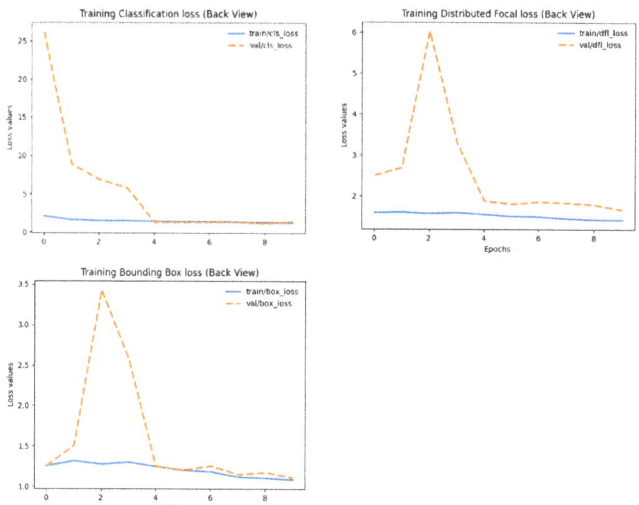

Figure 2. Training losses of the back-view model.

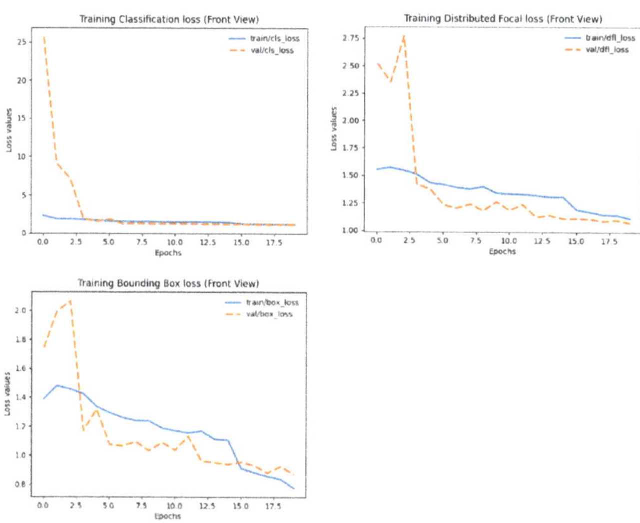

Figure 3. Training losses of the front-view model.

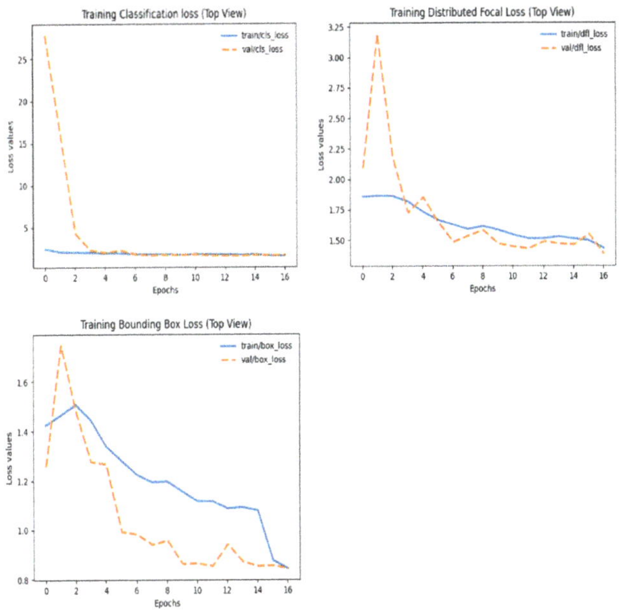

Figure 4. Training losses of the top-view model.

2.5. Data Analysis

A total of 1764 labeled images were used in this study, distributed across the three BCS classes (2.5, 3.0, and 3.5). The dataset was split into training (70%), validation (20%), and test (10%) sets using stratified sampling to maintain class balance. This resulted in 1234 images for training, 353 images for validation, and 177 images for testing. These partitions were fixed throughout all experiments to ensure consistent evaluation of model performance.

The trained model's performance was evaluated using three key assessment methods. First, precision/recall curves were employed to assess classification accuracy across BCS categories (2.5, 3.0, 3.5), offering insight into the balance between precision, which measures the correctness of positive predictions, and recall, which reflects the detection rate. These curves were calculated separately for each camera view, including the front, top, and back perspectives.

Next, confusion matrices were analyzed to assess misclassification rates and error patterns, helping to identify systematic misclassification trends. Overlapping feature representations between BCS classes were examined to determine where the model had difficulty distinguishing between categories.

Finally, Confidence Threshold Optimization was used to find the best probability cutoffs for classification performance. Thresholds were optimized separately for the front, top, and back views to enhance detection accuracy and provide more precise predictions.

Model training was performed on Google Colab using a GPU-accelerated environment with an NVIDIA Tesla T4 GPU and 16 GB RAM. After training, the model was tested on a local machine equipped with an Intel Core i7-12700H processor and 32 GB RAM, where CPU-based inference validation was conducted to evaluate computational performance in a real-world application.

The final annotated dataset included a total of 705 images across the three BCS classes after preprocessing and class adjustment: 210 images for BCS 2.5, 315 for BCS 3.0, and 180 for BCS 3.5. These were distributed approximately 35% from the front view, 40% from the rear view, and 25% from the top view. The confusion matrices and precision/recall (PR) curves were generated from a test set consisting of 10% of the entire dataset, maintaining proportional class representation. Apparent discrepancies between the confusion matrices and PR curves are due to differences in evaluation methods: the PR curves evaluate performance across multiple thresholds, while the confusion matrices show performance at a fixed confidence threshold (0.5). To improve clarity, we have included a supplementary table summarizing the number of images per class and dataset split (training, validation, test). This ensures complete traceability of classification performance across all BCS categories.

The overview of the research design includes (a) placement of 2D cameras around the AMS weighing station; (b) an example of cow ID and BCS tracking in an Excel spreadsheet; (c) the object detection annotation process in Roboflow, as shown in Figure 5; (d) sample annotated images from front, top, and back views; (e) dataset split and augmentation process; (f) YOLOv8 model training in Google Colab, as depicted in Figure 6; (g) a comparison of precision/recall curves for front, rear, and top views; and (h) as shown in Figure 7 a confusion matrix displaying misclassification patterns.

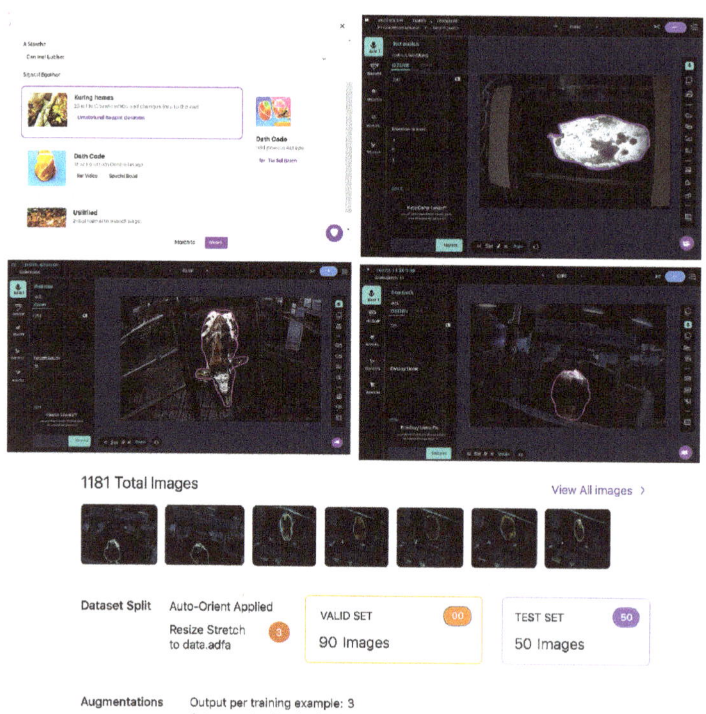

Figure 5. Roboflow annotations "(**top left**): Roboflow project initialization; (**top right,middle left,middle right**): SAM annotation tool; (**bottom center**): augmentations".

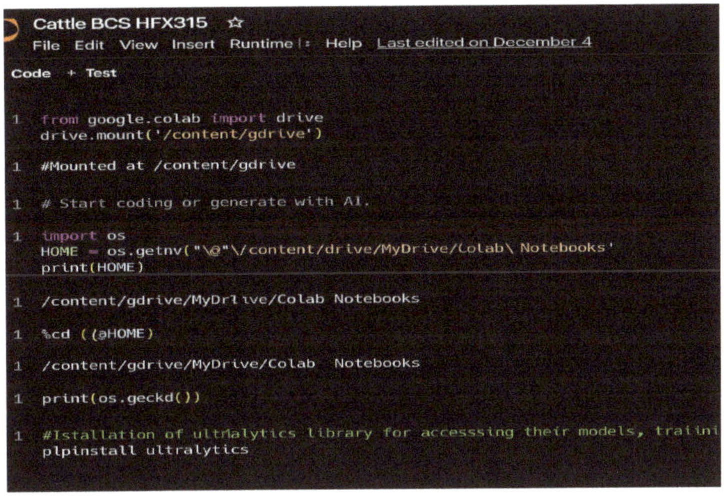

Figure 6. Google Colab.

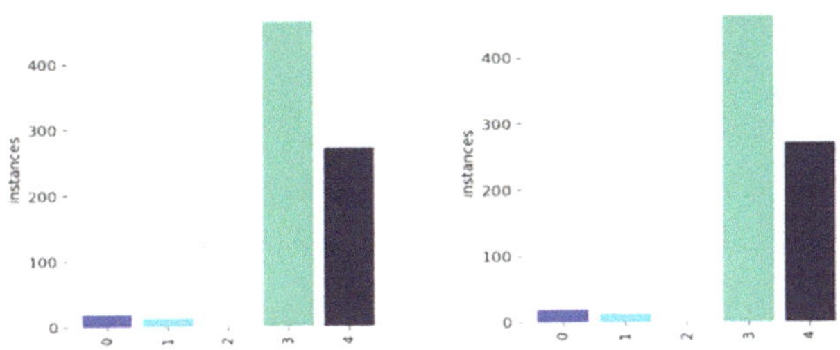

Figure 7. Class imbalance (**left**) and adjusted classes (**right**).

3. Results

3.1. Model Performance Based on Precision/Recall Curves

The precision/recall (PR) curve was used to evaluate the classification performance of the model across three different camera views, including front, top, and back.

The class-wise Average Precision (AP) was calculated for three body condition score (BCS) categories (2.5, 3.0, and 3.5), along with the mean Average Precision (mAP@0.5) across all classes.

For the front view, Class 2.5 achieved an AP of 0.519, Class 3.0 reached an AP of 0.518, and Class 3.5 had an AP of 0.260. This resulted in an overall mAP@0.5 of 0.432.

For the top view, Class 2.5 obtained an AP of 0.402, Class 3.0 an AP of 0.490, and Class 3.5 an AP of 0.372. The overall mAP@0.5 for this view was 0.421.

For the back view, Class 2.5 recorded an AP of 0.428, Class 3.0 an AP of 0.483, and Class 3.5 an AP of 0.439. This view led to the highest overall mAP@0.5 of 0.439, as shown in Figure 8.

The model achieved a mean Average Precision (mAP@0.5:0.95) of 0.41 on the validation set after 50 epochs. Class-specific mAPs were 0.35 for BCS 2.5, 0.47 for BCS 3.0, and 0.42 for BCS 3.5. These values show that the model generalizes reasonably well across all classes, with slightly higher precision for mid-range BCS scores.

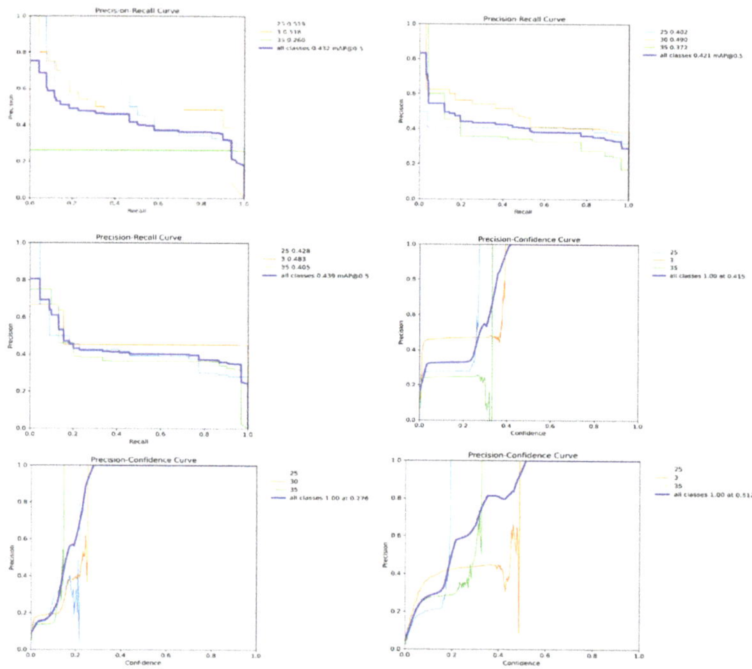

Figure 8. Precision/recall curves "(**top left**): front view; (**top right**): top view; (**middle left**): back view" and/precision/confidence curves "(**middle right**): front view; (**bottom left**): top view; (**bottom right**): back view".

To quantitatively support our conclusion that the top view performed the worst, we analyzed per-class accuracy and recall across all views (Table 1). The top view had the lowest precision and recall scores for all classes, with Class 3.0 achieving only 0.03 recall and Class 2.5 achieving 0.15. Similarly, Class 2.5 showed the lowest classification performance across all camera views, with high misclassification rates and significantly lower recall (e.g., 0.15 in the top view, 0.50 in the front view, and 0.52 in the back view). These metrics confirm the textual statements and highlight Class 2.5 as the most difficult to classify reliably due to feature overlap with background and neighboring BCS classes.

Table 1. Per-view and per-class classification performance metrics, including test sample counts, correct predictions, misclassifications, precision, and recall for BCS categories 2.5, 3.0, and 3.5.

BCS Class	View	Test Samples	Correct	Misclassified	Precision	Recall
2.5	Back	21	11	10	0.55	0.52
2.5	Front	20	10	10	0.50	0.50
2.5	Top	20	3	17	0.15	0.15

3.2. Precision/Confidence Analysis

The precision/confidence curves were created to evaluate the model's confidence at various probability thresholds for classification accuracy.

For the front view, the model achieved 100% precision at a confidence threshold of 0.415. Class 2.5 showed consistent precision growth, while Classes 3.0 and 3.5 experienced fluctuations in confidence scores (Figure 8).

In the top view, the optimal confidence threshold was 0.276, where precision achieved 100%. Classes 2.5 and 3.0 showed moderate improvements in precision, while Class 3.5 displayed greater variability.

In the back view, the model achieved 100% precision at a confidence threshold of 0.512. Classes 2.5 and 3.0 showed consistent improvements, but Class 3.5 experienced significant fluctuations in detection accuracy (Figure 8).

3.3. Classification Performance Based on Confusion Matrices

In Figure 9, the confusion matrices show classification accuracy and misclassification patterns across various BCS categories and camera views.

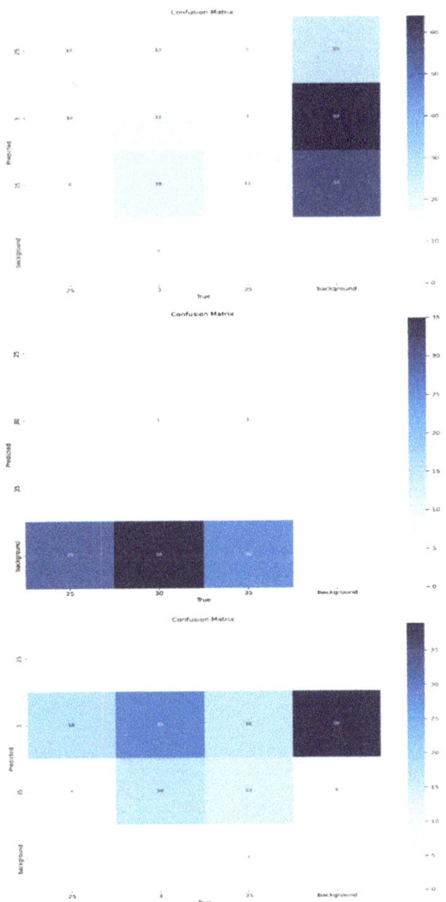

Figure 9. Confusion matrices ((**top**): front view; (**middle**): top view; (**bottom**): back view). Rows represent true labels, and columns represent predicted labels. Each cell shows the number of test samples per class. The back view exhibits the highest classification accuracy.

For the front view, Class 3.0 was accurately classified in 64 cases, while Class 3.5 was correctly identified in 57 cases. However, Class 2.5 had only 10 correct predictions, with 25 instances misclassified as background. Background data was often misclassified as Class 3.0, with 64 occurrences, indicating significant overlap in feature representation (Figure 9).

For the top view, a strong classification bias was observed, with nearly all predictions assigned to the background class. Class 2.5 was misclassified as background in 29 instances,

Class 3.0 in 35 instances, and Class 3.5 in 25 instances. Only one correct prediction was recorded for Class 3.0, indicating poor classification performance in this view (Figure 7). For the back view, Class 3.0 achieved 29 correct classifications. However, Class 2.5 was often misclassified as Class 3.0 in 18 instances, and background data was frequently misclassified as Class 3.0 in 39 instances. Class 3.5 had just 13 correct predictions, highlighting significant misclassification across categories.

Table 2 shows test set performance for each BCS class and view, including sample counts, correct predictions, misclassifications, precision, and recall. For example, while Class 2.5 in the back view had moderate average precision (AP = 0.428) on the PR curve, it also had a high number of misclassifications due to feature overlap with background and Class 3.0, especially at lower thresholds. This highlights the importance of interpreting PR and confusion matrix metrics together. The back view achieved higher overall mAP but still faced challenges with some boundary cases in Class 2.5.

Table 2. Classification performance per camera view and BCS class, including sample counts, correct predictions, misclassifications, precision, and recall.

View	BCS Class	Test Samples	Correct Predictions	Misclassifications	Precision	Recall
Back	2.5	21	11	10	0.55	0.52
	3.0	32	29	3	0.87	0.91
	3.5	18	13	5	0.72	0.72
Front	2.5	20	10	10	0.50	0.50
	3.0	30	24	6	0.80	0.80
	3.5	18	13	5	0.72	0.72
Top	2.5	20	3	17	0.15	0.15
	3.0	30	1	29	0.03	0.03
	3.5	18	2	16	0.11	0.11

4. Discussion

This study evaluated the effectiveness of a deep learning-based body condition scoring (BCS) system using 2D images captured from three camera perspectives: front, top, and back. The goal was to determine which view provided the most discriminative features for BCS classification and to identify the system's limitations for different classes. The results show that the back view consistently achieved the best performance, with the highest mAP and class-specific accuracy, especially for BCS 3.0. This is likely because the back view exposes the tailhead, hooks, and pins—areas known to be important for manual BCS scoring [17]. These anatomical landmarks are more visually accessible and less affected by lighting distortions or occlusion in the back view compared to the front or top.

In contrast, the top view produced the weakest performance, with significantly reduced precision and recall across all BCS categories. For example, Class 3.0 achieved a recall of only 0.03, and Class 2.5 achieved just 0.15. This suggests that top-down imaging may lack the visual granularity necessary to differentiate between subcutaneous fat layers and bony structures, which are essential for accurate BCS estimation. These results are consistent with previous studies, which have shown that top views are prone to perspective flattening and poor region localization in livestock applications [17].

The front view exhibited moderate performance, but it also faced significant class overlap—especially between Classes 2.5 and 3.0. This likely results from limited visibility of flank and pelvic landmarks from the front, which makes it harder for the model to identify distinguishing features. Additionally, cows in motion or poorly centered within the frame may have added to the prediction noise.

A consistent finding across all viewpoints was that Class 2.5 was the hardest to identify, with the highest rates of misclassification and the lowest precision and recall. This probably stems from both biological and imaging difficulties. Morphologically, Class 2.5 cows show subtle visual differences from nearby classes, and their lean body mass often merges with background or barn elements in 2D images. Confusion matrices and PR curves support this, especially in the top view, where Class 2.5 had a recall of just 0.15 [18]. Improving contrast, increasing annotation density, or using depth data could help address this.

These results have practical implications for on-farm deployment of automated BCS systems. Specifically, rear-mounted imaging appears to be the most effective strategy for capturing biologically informative views while minimizing classification errors. Integrating such a system into milking parlors, chutes, or alleyways would require minimal behavioral disruption. However, further improvements are needed to increase sensitivity to under-conditioned animals, which are often of greatest concern in herd management decisions [19,20].

From a technical perspective, we conducted three separate training trials to evaluate model stability. The results demonstrated consistent classification outcomes, with minimal variation in mAP and class-wise recall across runs (standard deviation < 0.015). Although this does not replace full cross-validation, it supports the reliability of the architecture given our dataset constraints. In future work, we plan to implement k-fold cross-validation, test generalizability across breeds and farm environments, and expand the BCS range to include classes 1.5 and 2.0 once enough data is available [21].

Compared to the existing literature, this study introduces a novel camera-perspective analysis using YOLOv8 and a simplified annotation framework, making it more practical for small- to mid-sized farms with limited infrastructure [22]. The focus on 2D RGB imaging—without needing depth or thermal sensors—shows that acceptable performance is possible even in resource-limited settings [23,24]. However, adding multi-modal sensors and temporal features (from video streams) could improve robustness in changing field conditions.

Overall, the results highlight the importance of camera placement and balanced datasets in AI-based animal monitoring. By tackling both technical and operational challenges of BCS classification, this work paves the way for scalable, cost-effective deployment of precision livestock technologies.

5. Conclusions

This study presents an innovative method for automated BCS classification that combines 2D imaging with deep learning. The strategic assessment of multiple camera angles sets this work apart from previous studies, which mainly relied on single views or expensive 3D systems. The results show that the rear view provides the most accurate capture of key BCS features, confirming that this cost-effective approach is practical for farm use.

By addressing the limitations of manual scoring and costly automated systems, this study lays the foundation for the wider adoption of precision livestock monitoring, especially among small and medium-sized dairy farms. Future research should focus on optimizing multi-view integration and improving model robustness to further increase classification accuracy. Additionally, expanding datasets, refining model architectures, and integrating complementary sensor technologies will strengthen the reliability and practicality of 2D-imaging-based BCS assessment in various farm environments. Efforts should also be made to make these systems more adaptable to small and medium-sized farms, ensuring that automated BCS monitoring can be easily implemented regardless of farm infrastructure. By incorporating these advancements, automated BCS systems can enhance herd health monitoring, boost productivity, and promote more sustainable dairy farming practices.

Author Contributions: T.K. and R.L. contributed to the conceptualization of the study; R.L. developed the methodology; T.K. was responsible for the software implementation and conducted the formal analysis; validation was carried out by T.L. and M.O.; investigation was undertaken by M.O.; resources were provided by R.L. and T.K.; data curation was performed by J.W.B., R.L. and T.K.; the original draft was prepared by R.L.; review and editing were conducted by M.O.; visualization was handled by M.O.; and supervision was provided by M.O. All authors have read and agreed to the published version of the manuscript.

Funding: This research received no external funding.

Data Availability Statement: The data generated and analyzed during this study will be made available upon reasonable request.

Acknowledgments: The authors express their gratitude to the Centre for Livestock Experiments (SHF) at the Norges Miljo-og Biovitenskaplige Universitet (NMBU), Norway, for their invaluable support in providing research facilities essential for the successful execution of this study. We also acknowledge the contributions of research staff and farm personnel who facilitated data collection and ensured the welfare of the animals involved in this study.

Conflicts of Interest: The authors declare no conflicts of interest.

Abbreviations

The following abbreviations are used in this manuscript:

AI	Artificial Intelligence
AMS	Automated Milking System
AP	Average Precision
BCS	Body Condition Score
Colab	Google Colaboratory
RFID	Radio-Frequency Identification
SD	Secure Digital (card)

References

1. Lemmens, L.; Schodl, K.; Fuerst-Waltl, B.; Schwarzenbacher, H.; Egger-Danner, C.; Linke, K.; Suntinger, M.; Phelan, M.; Mayerhofer, M.; Steininger, F.; et al. The Combined Use of Automated Milking System and Sensor Data to Improve Detection of Mild Lameness in Dairy Cattle. *Animals* **2023**, *13*, 1180. [CrossRef] [PubMed]
2. Roche, J.R.; Friggens, N.C.; Kay, J.K.; Fisher, M.W.; Stafford, K.J.; Berry, D.P. Body condition score and its association with dairy cow productivity, health, and welfare. *J. Dairy Sci.* **2009**, *92*, 5769–5801. [CrossRef] [PubMed]
3. Bewley, J.M. Automated Body Condition Scoring of Dairy Cattle: Technical and Economic Feasibility. Doctoral Dissertation, Purdue University, West Lafayette, IN, USA, 2008.
4. Krukowski, M.; McKague, K.; Valerio, A. Limitations of manual BCS evaluation in dairy farms: A systematic review. *Appl. Anim. Behav. Sci.* **2021**, *241*, 105378. [CrossRef]
5. Li, X.; Lv, C.; Wang, W.; Li, G.; Yang, L.; Yang, J. Generalized focal loss: Towards efficient representation learning for dense object detection. *IEEE Trans. Pattern Anal. Mach. Intell.* **2023**, *45*, 3139–3153. [CrossRef] [PubMed]
6. López-Gatius, F.; Yániz, J.; Madriles-Helm, D. Effects of body condition score and score change on the reproductive performance of dairy cows: A meta-analysis. *Theriogenology* **2003**, *59*, 801–812. [CrossRef] [PubMed]
7. Halachmi, I.; Klopčič, M.; Polak, P.; Roberts, D.J.; Bewley, J.M. Automatic assessment of dairy cow body condition score using 3D image analysis. *J. Dairy Sci.* **2019**, *96*, 8047–8059. [CrossRef]
8. Mottram, T.T.F.; den Uijl, I. Health and welfare monitoring of dairy cows. In *Digital Agritechnology: Applications and Future Prospects*; Khan, M.I.R., Jamal, M.M., Eds.; Elsevier: Amsterdam, The Netherlands, 2022; pp. 113–142. [CrossRef]
9. Simitzis, P.; Tzanidakis, C.; Tzamaloukas, O.; Sossidou, E. Contribution of Precision Livestock Farming Systems to the Improvement of Welfare Status and Productivity of Dairy Animals. *Dairy* **2022**, *3*, 12–28. [CrossRef]

10. Zhao, K.; Zhang, M.; Shen, W.; Liu, X.; Ji, J.; Dai, B.; Zhang, R. Automatic body condition scoring for dairy cows based on efficient net and convex hull features of point clouds. *Comput. Electron. Agric.* **2023**, *205*, 107588. [CrossRef]
11. Islam, A.; Lomax, S.; Doughty, A.; Islam, M.R.; Jay, O.; Thomson, P.; Clark, C. Automated monitoring of cattle heat stress and its mitigation. *Front. Anim. Sci.* **2021**, *2*, 737213. [CrossRef]
12. Mottram, T.; den Uijl, I. Integrating AI into dairy farming: A future perspective. *Anim. Sci. J.* **2022**, *93*, 13671. [CrossRef]
13. Ministry of Agriculture and Food. *Regulations on the Keeping of Cattle*; Norwegian Ministry of Agriculture and Food: Oslo, Norway, 2007. Available online: https://nibio.brage.unit.no/nibio-xmlui/bitstream/handle/11250/2494661/NILF-Diskusjonsnotat-2007-02.pdf?isAllowed=y&sequence=2&utm (accessed on 18 May 2025).
14. Ministry of Agriculture and Food. *Regulations Relating to the Use of Animals in Research*; Norwegian Ministry of Agriculture and Food: Oslo, Norway, 2015. Available online: https://www.regjeringen.no/no/dokumenter/forskrift-om-bruk-av-dyr-i-forsok/id2425378/ (accessed on 18 May 2025).
15. Lewis, L. *Overview of the Barn and Camera Placement*; Original Figure, Created by the Author During Data Collection at the Animal Research Centre; NMBU: Ås, Norway, 2024.
16. Siachos, N.; Lennox, M.; Anagnostopoulos, A.; Griffiths, B.E.; Neary, J.M.; Smith, R.F.; Oikonomou, G. Development and validation of a fully automated 2-dimensional imaging system generating body condition scores for dairy cows using machine learning. *J. Dairy Sci.* **2024**, *107*, 2499–2511. [CrossRef] [PubMed]
17. Van Nuffel, A.; Zwertvaegher, I.; Van Weyenberg, S.; Pastell, M.; Thorup, V.M.; Bahr, C.; Sonck, B.; Saeys, W. Lameness detection in dairy cows: Part 2. Use of sensors to automatically register changes in locomotion or behavior. *Animals* **2015**, *5*, 861–885. [CrossRef] [PubMed]
18. Biscarini, F.; Nicolazzi, E.L.; Stella, A.; Boettcher, P.J.; Gandini, G. Challenges and opportunities in genetic improvement of local livestock breeds. *Front. Genet.* **2015**, *6*, 33. [CrossRef] [PubMed]
19. Neethirajan, S. Transforming the adaptation physiology of farm animals through sensors. *Animals* **2020**, *10*, 1512. [CrossRef] [PubMed]
20. Zheng, Z.; Wang, P.; Liu, W.; Li, J.; Ye, R.; Ren, D. Distance-IoU loss: Faster and better learning for bounding box regression. *Proc. AAAI Conf. Artif. Intell.* **2020**, *34*, 12993–13000. [CrossRef]
21. Wurtz, K.; Camerlink, I.; D'Eath, R.B.; Fernández, A.P.; Norton, T.; Steibel, J.; Siegford, J.; Raboisson, D. Recording behaviour of indoor-housed farm animals automatically using machine vision technology: A systematic review. *PLoS ONE* **2019**, *14*, e0226669. [CrossRef] [PubMed]
22. Yaseen, M. What is YOLOv8: An in-depth exploration of the internal features of the next-generation object detector. *arXiv* **2024**, arXiv:2408.15857. [CrossRef]
23. Summerfield, G.I.; De Freitas, A.; van Marle-Koster, E.; Myburgh, H.C. Automated Cow Body Condition Scoring Using Multiple 3D Cameras and Convolutional Neural Networks. *Sensors* **2023**, *23*, 9051. [CrossRef] [PubMed]
24. Zhang, Y.; Zhang, Q.; Zhang, L.; Li, J.; Li, M.; Liu, Y.; Shi, Y. Progress of machine vision technologies in intelligent dairy farming. *Appl. Sci.* **2023**, *13*, 7052. [CrossRef]

AgriEngineering

MDPI

Article

Automated Detection of Kinky Back in Broiler Chickens Using Optimized Deep Learning Techniques

Ramesh Bahadur Bist [1,2], Andi Asnayanti [3], Anh Dang Trieu Do [3], Yang Tian [1], Chaitanya Pallerla [1,4], Dongyi Wang [1,4,*] and Adnan A. K. Alrubaye [3,*]

[1] Department of Biological and Agricultural Engineering, University of Arkansas, Fayetteville, AR 72701, USA; rbbist@ncsu.edu (R.B.B.)
[2] Department of Biological and Agricultural Engineering, North Carolina State University, Raleigh, NC 27606, USA
[3] Department of Poultry Science, University of Arkansas, Fayetteville, AR 72701, USA; aasnayan@uark.edu (A.A.); ad086@uark.edu (A.D.T.D.)
[4] Department of Food Science, University of Arkansas, Fayetteville, AR 72701, USA
* Correspondence: dongyiw@uark.edu (D.W.); aakhalaf@uark.edu (A.A.K.A.)

Abstract: The global poultry industry faces growing challenges from skeletal disorders, with Kinky Back (KB) significantly impacting broiler welfare and production. KB causes spinal deformities that reduce mobility, feed access, and increase mortality. It often remains undetected in early subclinical stages. Traditional KB diagnosis methods are slow and subjective, and highlighting the need for an automated and objective detection. This study develops a machine learning approach for detecting KB in broilers using image data. Male Cobb 500 broilers were raised under controlled conditions and monitored over 7 weeks using overhead 4K video cameras. Behavioral and posture data related to KB were collected and annotated from images extracted from the videos. First, various optimizers (SGD, Adam, AdamW), image sizes, and data augmentation techniques were compared, and the best-performing optimizer, image size, and data augmentation technique were identified. These findings were then used to compare different original lightweight YOLO models trained and to identify the best models with further modifications to these configurations, aiming to improve detection accuracy. Different machine vision models were evaluated using precision, recall, F1-score, and mean average precision metrics to identify the best-performing approach. Among the tested optimizers, SGD achieved the highest precision (100%) and mAP_0.50–0.95 (74.7%), indicating superior localization and lower false-positive rates, while AdamW produced the highest recall (98.9%) with slightly lower precision. Image input size of 960 × 960 pixels yielded the best balance of precision (99.0%), recall (99.4%), and F1-score (99.2%). Data augmentation improved recall and reduced false negatives by confirming its value in enhancing model generalization. Among YOLO architectures, YOLOv9 performs best. Furthermore, the optimized YOLOv9 model, combined with augmentation and 960-sized images, achieved the highest performance, with a precision of 99.1%, recall of 100%, F1-score of 99.5%, and mAP of 78.0%. Overall, the proposed optimized YOLOv9-based system provides a reliable and scalable framework for automated detection of Kinky Back, supporting data-driven welfare management in modern poultry production.

Keywords: Kinky Back; machine learning; YOLO; optimizers; image size; augmentation; poultry

1. Introduction

The global poultry industry has experienced significant growth over the past few decades, primarily driven by rising demand for affordable animal protein [1,2]. This surge has been further supported by technological innovations and improved farming techniques that enhance production efficiency. Advances in genetic selection, nutrition, and management practices have contributed to the development of fast-growing broiler chickens [3]. Compared with chicken meat in the 1950s, modern broilers have doubled body mass in half the production time [4]. However, this rapid growth has inadvertently introduced health concerns that negatively affect animal health, productivity, and welfare, including myopathies (woody breast, spaghetti meat) [5,6] and skeletal issues, such as lameness [7,8]. Among skeletal disorders affecting modern broilers, spondylolisthesis is particularly concerning [9]. Spondylolisthesis is a physical disorder that is commonly referred to as a Kinky Back (KB). In addition, KB symptoms can be a manifestation of bacterial infection of the fourth thoracic vertebra (T4), known as spondylitis [10–12]. KB has garnered increasing attention due to its impact on broiler welfare and production efficiency (Figure 1). It is characterized by deformities of the thoracic vertebrae, leading to spinal cord compression, gait abnormalities, and, in some cases, paralysis [13]. Affected birds often struggle to access feed and water, contributing to poor performance and increased mortality [14].

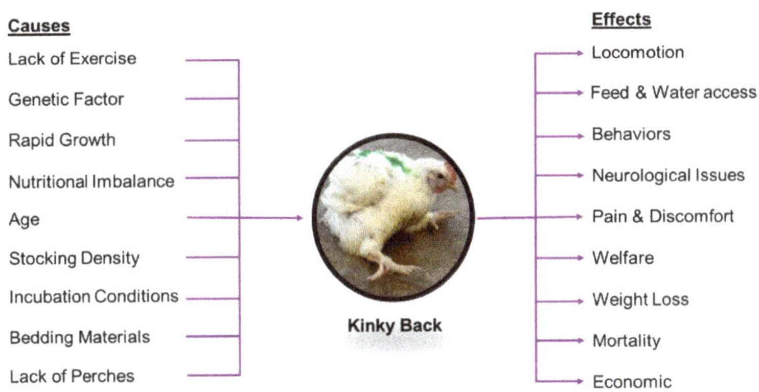

Figure 1. Causal Factors and Effects of Kinky Back in Broilers.

KB was first identified in Great Britain and subsequently reported in countries like Australia, Canada, the USA, and Germany [15–17]. It has been recognized as a global concern in the broiler production industry. The condition most frequently affects the fifth and sixth thoracic vertebrae between 3 and 6 weeks of age [9,18,19]. Clinical cases may affect only about 2% of the flock, and up to 60% of birds can be subclinically impacted, experiencing discomfort and mobility issues without visible symptoms [9,17]. This hidden prevalence complicates early intervention, as affected birds often go unnoticed until the condition worsens. In addition, detecting the early subclinical stage is difficult with standard observation methods. Difficulty in early detection underscores the need for advanced diagnostic tools to identify at-risk birds before symptoms progress.

Genetic predisposition is widely acknowledged as a primary factor of spondylolisthesis KB, with notable differences in prevalence across commercial hybrids [17]. Genetic differences influence bone and spinal risks, requiring tailored management. Additionally, dietary factors, such as high-energy and low-fiber diets, have been implicated in exacerbating spinal stress during early growth phases [18]. However, practical limitations often hinder the application of nutritional interventions in commercial systems. Whereas spondylitis KB is highly associated with opportunistic bacteria, particularly Enterococcus. Enterococcal spondylitis leads to a symmetrical hind limb paralysis as a consequence of the *Enterococcus cecorum* colonization of the free thoracic vertebrae [20,21].

To address the KB condition from both welfare and productivity perspectives, researchers have investigated ways to manage it through environmental enrichment. These approaches focus on promoting movement and strengthening musculoskeletal health [22]. Enrichment tools such as elevated platforms, straw bales, and interactive laser light projections have shown promising results in mitigating the incidence of KB [22]. These tools engage animals in dynamic behaviors, encouraging natural movement and exploration. By promoting increased activity, these strategies help reduce mechanical stress on the developing bones and spine. For instance, perching platforms and laser stimulation have been reported to decrease the prevalence of subclinical spondylolisthesis from as high as 60% to as low as 21% and 29%, respectively [19,22]. While these strategies offer preventive benefits, they are insufficient on their own due to the multifactorial etiology of the condition and the delayed detection of current methods. To address this complexity, integrating proactive monitoring with targeted interventions is essential for optimizing outcomes.

Historically, KB has relied on manual assessment of gait, posture, and activity levels [19,23]. These methods are inherently time-consuming, subjective, inconsistency, and vastly inadequate for identifying subclinical cases. Therefore, there is a critical need for automated tools with real-time detection capabilities and objectivity to identify indicators of the condition. The rise in precision livestock farming presents new opportunities to meet this need. Technologies such as computer vision and sensor-based monitoring pave the way for real-time health assessments [24]. Machine learning offers a powerful approach for analyzing behavioral and physiological data [25–27]. It enables accurate and consistent detection of health issues, including musculoskeletal disorders [28,29].

Recent advances in ML applications in livestock systems have demonstrated the utility of convolutional neural networks (CNNs), support vector machines (SVMs), and random forests for health monitoring [30–33]. In poultry production, these models have successfully identified lameness [34], footpad dermatitis [24], and mortality [35,36] through the image and sensor. Utilizing video footage and time-series data from activity monitoring enables the detection and classification of abnormal patterns [37,38]. However, these technologies have shown promise in other areas. Their application in detecting KB in broilers remains underexplored. Subtle behavioral signs, such as reduced perching, altered posture, or restricted movement, in the early stages can be challenging to notice. Machine learning offers a precise framework for identifying and analyzing these indicators.

This study aims to develop a robust, automated, and scalable machine learning-based approach for detecting KB once its symptoms are present. Detecting KB will help identify increasing cases of KB and facilitate removal, reducing pain and negative impacts on behaviors and welfare. The proposed research will collect image-based data during growth phases, including expert-guided labeling of clinical and subclinical cases, and train multiple machine learning models to assess detection performance. Ultimately, this research aims to develop an improved machine learning model to detect KB and help farmers make timely interventions. By addressing the current limitations in detection, the study has the potential to significantly enhance welfare and health management strategies within commercial poultry operations.

2. Materials and Methods

The materials and methods section clearly outlines how image data were collected, processed, and used to train a model for Kinky Back detection. It details animal housing and management, image collection and processing, data augmentation, the software and hardware used, the architecture, and evaluation metrics.

2.1. Broiler Housing and Management

KB cases were collected during the experimental lameness trial in Fall 2024. These experimental lameness procedures were conducted in accordance with the ethical standards approved by the Institutional Animal Care and Use Committee (IACUC) at the University of Arkansas (Protocol #23014). The study was conducted over 56 days at the University of Arkansas Poultry Research Farm Complex. A total of 720 male Cobb 500 chicks were randomly distributed into 12 pens. Two pens featured wire flooring as an incubation for lameness diseases, including KB, and ten had litter flooring for study treatments. Each pen initially housed 60 chicks, which were reduced to 50 per pen on day 14 (D14) to maintain an appropriate stocking density. In the last two weeks, we assigned two more pens with 10 broilers each, one of which has a KB for data collection. All pens were identical in size, measuring 1.5 m by 3.0 m.

Environmental conditions within the house were automatically controlled to maintain thermoneutral temperatures appropriate for broiler development. Temperature of 32 °C (90 °F) from days 1 to 3, 31 °C (88 °F) from days 4 to 6, 29 °C (85 °F) from days 7 to 10, 26 °C (80 °F) from days 11 to 14, and 24 °C (75 °F) thereafter. A constant photoperiod of 23 h light and 1 h dark (23L:1D) was maintained throughout the trial. Birds were provided with unrestricted access to clean water and fed industry-standard diets. Broilers receive starter crumbles from day 0 to 21 and grower pellets from day 22 to 56.

2.2. Video Recording and Data Collection

To monitor broiler behavior and detect signs of KB, a continuous video surveillance system was deployed throughout the 7-week trial. In this experiment, the LOREX Pro Series 4K 32-Camera, 8TB Wired NVR System (N884A64B, Lorex Corporation, Columbia, MD, USA) with A14 IP Bullet Cameras (E842CA, Lorex Corporation, Columbia, MD, USA) were used (Figure 2). Overhead cameras were installed in 8 selected pens to capture high-resolution (4K, 20 fps) video footage 24 h a day. From week 3 onward, birds were monitored both via video recording and in person for indicators of KB. This study focused on identifying abnormal postures, including sitting on the back or tail (particularly the "dog-sitting" position), walking backward, reduced mobility, uncoordinated gait, hock-sitting posture with extended legs, frequent sitting or lying, asymmetric leg movement, and difficulty maintaining balance [39,40] (Figure 3). After data collection, birds with KB were humanely euthanized.

Figure 2. Experimental pen layout and camera setup for data collection.

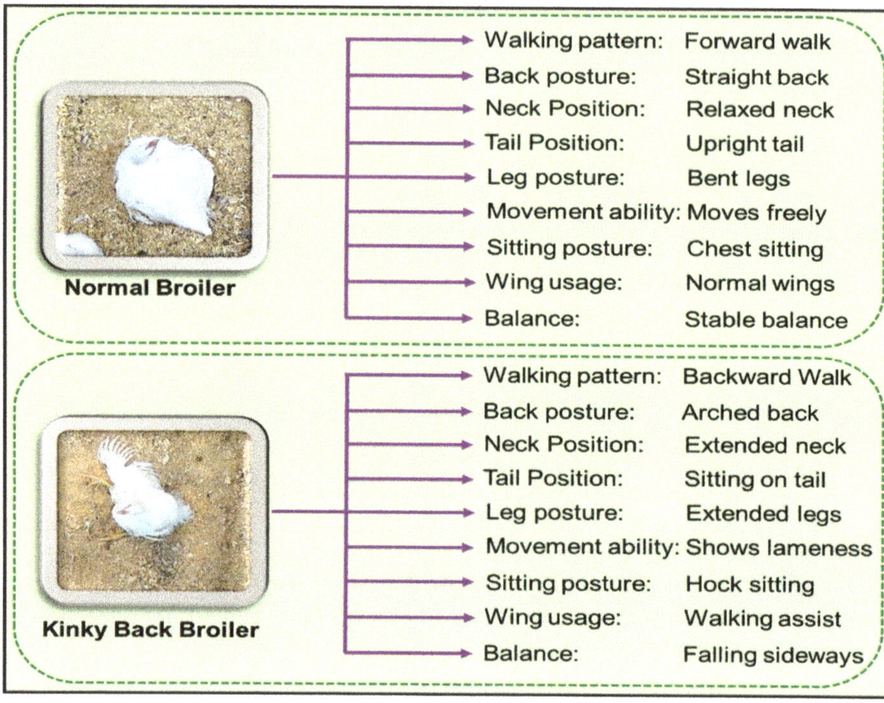

Figure 3. Physical differences between normal broiler and Kinky Back broiler.

2.3. Data Annotation and Labeling

The videos were randomly converted to images (2766 × 2082 pixels) using Python 3.13.0 code, and the images were then sorted based on KB presence. The 600-image datasets (5400 regular and 600 KB broiler instances) were split randomly into training (60%), validation (20%), and testing (20%) before labeling. For labeling those images, trained researchers manually annotated the images using Afpha Make Sense (www.makesense.ai, accessed on 1 October 2025). Broilers with KB in each image were labeled using bounding boxes and categorized based on behavioral and physical appearance.

2.4. Machine Learning Model Development

Several open-source deep learning models were initialized in their default configurations to establish baseline performance. Improving model accuracy and robustness required careful selection of optimizers and augmentation techniques during training (Figure 4). This study evaluated various optimizers (Stochastic Gradient Descent, Adam, and AdamW) and augmentation techniques (image size, stronger color, and aggressive augmentations) to determine the most effective combination for improving model performance. These optimizers were used to understand how each optimizer affected convergence, learning stability, and final accuracy [41]. For instance, Adam and AdamW adapt the learning rate during training, potentially improving convergence on complex datasets [42,43], whereas SGD often leads to better generalization [44,45]. In addition, various data augmentation techniques were also applied, including resizing input images (640, 960, 1280), enhancing color variations (hsv_h = 0.02; hsv_s = 0.8; hsv_v = 0.5), and using stronger augmentations such as mosaic, mix-up, flipping, rotation, and brightness adjustment. These augmentations enhance data diversity, enabling the model to generalize more effectively and avoid overfitting [46]. In addition, the reason for using different image sizes is that YOLO models usually use a 640 image size and other arguments by default. Therefore, this study wants to compare the default setting with other settings. Based on prior findings, improved performance is achieved with increased training epochs of 200 and a batch size of 16 [47].

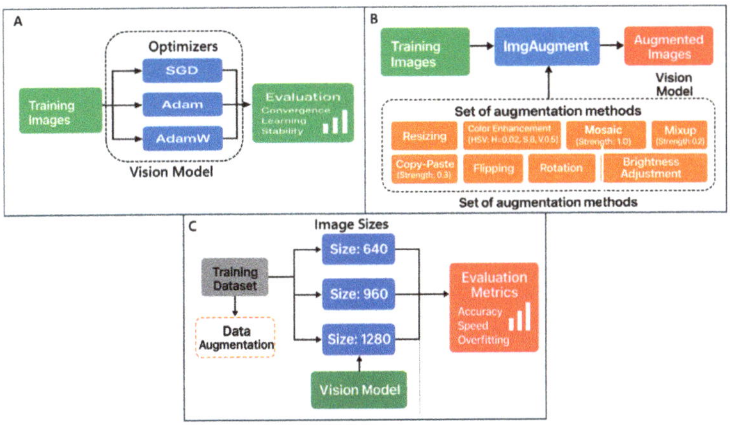

Figure 4. Modular (**A**) Optimizer, (**B**) Data Augmentation, and (**C**) Image Size Augmentation Framework for Robust Vision Model Training.

Furthermore, this study utilized various small-sized, lightweight YOLO models (YOLOv9s, YOLOv10s, YOLO11s, and YOLO12s) for comparison, identifying the most effective YOLO models [48]. These models were selected for their reduced computational requirements and enhanced accuracy, making them suitable for real-time applications. Each YOLO variant has unique architectural differences, including modifications to backbone, neck, and detection head structures [48]. After evaluating all combinations, the best-performing YOLO model was retrained using the most effective optimizer and augmentation setup to ensure maximum performance. The goal was to identify the most accurate and efficient model for further fine-tuning and deployment. To train these models, the high-performance computing system (Arkansas High Performance Computing Center, Fayetteville, AR, USA) was utilized, with training conducted in Jupyter Notebook 4.2 using NVIDIA A100 GPUs and 64 CPU cores.

2.5. Model Evaluation and Statistical Analysis

Performance metrics, including precision, recall, F1-score, and mean average precision (mAP), were used for evaluation across all experiments [47,48]. Precision quantified the proportion of correctly identified positive detections, while recall measured the model's ability to detect true positive instances. The F1-score provided a balanced metric combining precision and recall. Mean Average Precision evaluated detection accuracy across varying intersection-over-union (IoU) thresholds. These metrics helped spot common misclassification patterns and compare models. Each YOLO configuration was trained and validated separately with the same dataset split, and results were compared using these metrics. Furthermore, each YOLO variant had its own architecture and optimization strategy, so they were benchmarked independently.

3. Results and Discussions

3.1. Effect of Optimization Parameters and Preprocessing Techniques

3.1.1. Optimization Parameter Comparison

This study compares the performance of three optimizers (Adam, AdamW, and SGD) for detecting KB using a YOLO-based deep learning model. The results showed that SGD performed the best overall as shown in Table 1 and Figure 5. It achieved the highest precision, mAP_0.50, and mAP_0.50–0.95 among the three. While AdamW had slightly higher recall and F1-score, SGD showed stronger overall detection accuracy and localization. High precision means the model made fewer false detections, which is essential in real-time applications. High precision is critical in commercial settings to avoid unnecessary inspections, reduce labor costs, and prevent alert fatigue. A higher mAP_0.50–0.95 also suggests that the model trained with SGD was better at identifying and localizing different variations in the target condition. Previous studies in poultry detection mostly used Adam, AdamW and SGD optimizers [49]. Using SGD reported significantly higher generalization than Adam [49,50]. Figure 4 shows similar results, with the SGD optimizer outperforming the Adam optimizer. The SGD optimizer consistently demonstrated improved performance, whereas other optimizers showed fluctuating trends and less stable convergence. Model performance improves as the number of epochs increases from epoch 0 to 200. This shows the importance of selecting the right optimizer based on the detection goal.

Table 1. Validation performance of various optimizers for Kinky Back detection (%).

Optimizers	Precision	Recall	mAP_0.50	mAP_0.50–0.95	F1-Score
Adam	86.2	76.5	89.1	64.9	81.0
AdamW	93.6	98.1	96.6	67.8	96.0
SGD	99.1	89.9	97.5	74.7	95.0

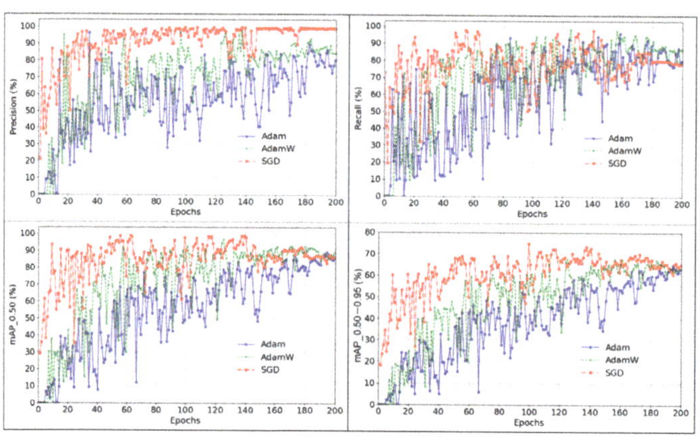

Figure 5. Comparison of Optimizers' Performance for Detecting Kinky Back.

A similar trend was observed in the test dataset (Table 2). SGD achieved 100% precision, demonstrating its strong ability to avoid false positives. F1-score was also the highest at 95.6%, confirming its balanced performance. However, the recall was slightly lower at 91.6%, resulting in a higher false-negative rate (FNR) of 8.4%. In comparison, AdamW achieved the highest recall at 98.9% and the lowest FNR of 1.1%, indicating that it was best at detecting all actual cases. Adam had the lowest F1-score at 92.0% and a higher FNR than AdamW. Overall results confirm that SGD is most effective when precision is critical, while AdamW is useful when high recall is required. However, Adam generally provides greater training stability due to its adaptive learning rate and moment estimates.

Table 2. Performance comparison of optimizers on the test dataset for Kinky Back detection (%).

Optimizers	Precision	Recall	F1-Score	FNR
Adam	87.8	96.6	92.0	3.4
AdamW	88.5	98.9	93.4	1.1
SGD	100.0	91.6	95.6	8.4

3.1.2. Evaluation-Based Image Size and Augmentation

This study compares the effect of different image sizes (640, 960, and 1280 pixels) on model performance for detecting KB in broilers. Different image sizes are selected by changing the image size input during model training. The model trained on 960-sized images achieved the best balance between precision, recall, and mAP. Similarly, 960-sized image achieved 98.3% precision, 97.7% recall, 98.7% mAP_0.50, and 76.3% mAP_0.50–0.95, with the highest F1-score of 98.0% (Table 3). This indicates strong detection ability with minimal false positives and negatives. While 1280-sized images yielded a slightly higher mAP_0.50–0.95 of 77.3%, they required more training time and showed a slight drop in precision. Larger sizes slowed inference and increased the risk of overfitting, making them less practical for real-time detection in commercial settings. On the other hand, the 640-sized images trained faster at 14.7 min but had lower recall and F1-score. More training time requires additional computational resources, such as more GPUs, thereby increasing computational costs [51]. Small-sized images train faster with deep learning models than large images [52]. However, large-sized images do not perform as well as medium-sized images. Therefore, the results suggest that using medium-sized images offers the best trade-off between speed and performance without significantly increasing computational cost in real-time systems.

Table 3. Effect of image size and augmentation on model performance (%) and training time (min) for Kinky Back detection.

Configurations	Type	Precision	Recall	mAP_0.50	mAP_0.50–0.95	F1-Score	Training Time Taken
Image Sized	Image 640	98.1	85.7	98.2	73.6	93.0	14.7
	Image 960	98.3	97.7	98.7	76.3	98.0	22.7
	Image 1280	95.8	95.9	98.2	77.3	96.0	35.0
Augmentation	With	97.5	91.6	98.1	73.2	94.5	15.6
	Without	98.1	85.7	98.2	73.6	91.6	15.2

In this study, there were fewer KB samples than normal broilers, so we used data augmentation to balance the dataset. Data augmentation methods mitigate overfitting issues in computer vision and machine learning [53]. Therefore, this study also compares the impact of data augmentation on model performance (Table 3). When data augmentation was applied, the model achieved 97.5% precision and 91.6% recall with an F1-score of 94.5% and mAP_0.50–0.95 of 73.2%. Without augmentation, precision remained similar at 98.1%, but recall dropped significantly to 85.7%. The result suggests that augmentation enhanced the model's ability to generalize, which leads to better detection of true classes. Similar results were observed in the previous study [54]. Data augmentation is effective in tasks like poultry behavior detection and helps models avoid overfitting [27]. These results demonstrate that even small changes from data augmentation can lead to noticeable improvements in recall and overall model robustness.

Table 4 compares the effects of image size and data augmentation on test performance for KB detection using a deep learning model. Among different image sizes, 960 pixels achieved the best results, with a precision of 99.0%, a recall of 99.4%, and the highest F1-score of 99.2%. The performance for both validation and test was higher for image size 960. The 640-pixel size had slightly higher precision (99.4%) but lower recall (93.9%) and a higher FNR of 6.1%. In contrast, 1280 pixels showed lower precision (93.6%) despite achieving a reasonable recall rate (97.8%). These results suggest that 960-pixel images offer the best balance of accuracy and sensitivity without excessive false positives or longer training times. The findings show that image size and augmentation are key to achieving both high precision and recall.

Table 4. Comparison of image size and augmentation effects on test dataset performance (%) for Kinky Back detection.

Preprocessing	Type	Precision	Recall	F1-Score	FNR
Image Sized	Image 640	99.4	93.9	96.6	6.1
	Image 960	99.0	99.4	99.2	0.6
	Image 1280	93.6	97.8	95.6	2.2
Augmentation	With	99.4	93.9	96.6	6.1
	Without	100.0	87.7	93.5	12.3

3.2. Comparison of YOLO Architectures with Optimized Settings

This study compares several small YOLO model architectures, including YOLOv9, YOLOv10, YOLO11, and YOLO12. Among the base models, YOLOv9 demonstrated high precision (98.1%) but moderate recall (85.7%), with an F1-score of 91.6% (Table 5). YOLO11 had the highest recall at 94.1% and a strong F1-score of 93.6%. However, YOLOv10 and YOLO12 performed slightly lower in both precision and recall. Given the YOLOv9 model's higher performance, this study focused on further improving It by applying various optimizations and data augmentation techniques based on the above findings.

Table 5. Validation performance comparison of YOLO architectures and optimized YOLOv9 models with augmentation strategies for Kinky Back detection (%).

Models	Precision	Recall	mAP_0.50	mAP_0.50–0.95	F1-Score
YOLOv9	98.1	85.7	98.2	73.6	91.6
YOLOv10	97.1	83.6	91.5	67.8	89.8
YOLO11	93.1	94.1	97.4	72.5	93.6
YOLO12	91.5	89.9	93.6	71.5	90.7
YOLOv9-Optimizer + Augmentation	97.9	100.0	98.8	73.8	98.9
YOLOv9-Optimizer + I960 + Augmentation	99.1	100.0	98.9	78.0	99.5

The YOLOv9 model, trained with SGD and augmentation, achieved perfect recall (100%) and maintained high precision (97.9%). When combined with 960-pixel image input and augmentation (YOLOv9-Optimized + I960 + Augmentation), the model achieved the highest precision of 99.1%, perfect recall, and the best F1-score of 99.5%, along with an improved mAP_0.50–0.95 of 78.0%. These results highlight that optimization and proper preprocessing can greatly enhance detection accuracy. Overall, YOLO versions that initially showed lower recall or precision benefited from these optimizations.

From Table 6, YOLOv9 achieved 100% precision but a lower recall of 87.7%, resulting in an F1-score of 93.5% and an FNR of 12.3%. Similar trends were observed in YOLOv10 and YOLO12, both of which exhibited high precision but recall rates below 88%. YOLO11 differed by achieving very high recall (99.4%) and a low FNR (0.6%), but its precision dropped to 86.4%. These base models highlight a trade-off between precision and recall, affecting overall detection reliability. Similarly, optimized versions of YOLOv9 were tested with augmentation and an input image size of 960 pixels. These improvements demonstrate that optimization and preprocessing significantly enhance detection balance and reduce the number of missed cases. Our results align with studies that emphasize data augmentation and input size adjustments as key factors for boosting model robustness (Figure 6). This comparison confirms that customized YOLOv9 models provide superior and more reliable detection of KB. Thus, this study successfully developed an improved machine learning model to detect KB, which can help farmers in the future identify KB cases on the farm and develop ways to mitigate them.

Table 6. Test dataset performance comparison of YOLO architectures and optimized YOLOv9 models for Kinky Back detection (%).

Models	Precision	Recall	F1-Score	FNR
YOLOv9	100.0	87.7	93.5	12.3
YOLOv10	98.7	87.7	92.9	12.3
YOLO11	86.4	99.4	92.5	0.6
YOLO12	100.0	87.2	93.1	12.8
YOLOv9-Optimizer + Augmentation	99.4	95.5	97.4	4.5
YOLOv9-Optimizer + I960 + Augmentation	98.9	97.8	98.3	2.2

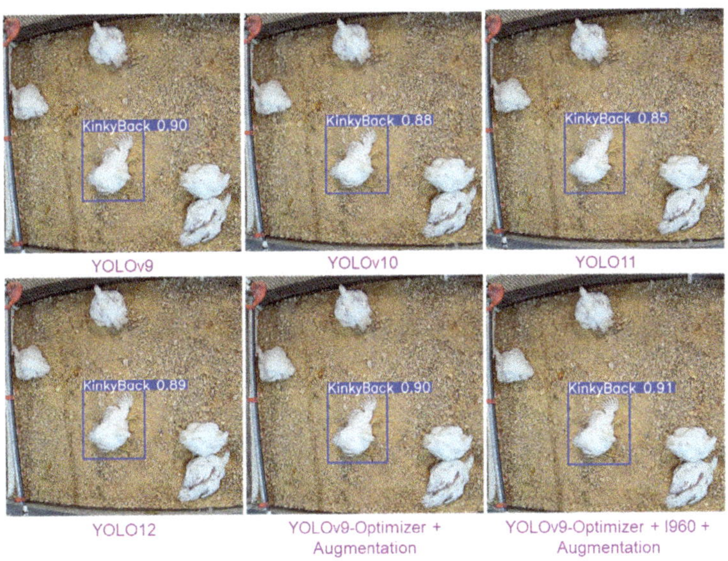

Figure 6. Prediction results of different models with bounding boxes and confidence scores.

3.3. Limitation and Future Direction

Despite the machine learning model's promising performance in detecting KB, several limitations were identified. One key challenge is the visual similarity between affected and healthy birds in static images at an early stage. It can lead to misclassifications where subtle deformities are not yet apparent. In addition, the model's efficacy is constrained by its training on data from specific housing systems. It may limit its applicability to settings such as free-range or enriched cages, where lighting, crowding, and background conditions vary significantly. Moreover, limited datasets are due to fewer KB cases than normal broilers, and a single class may introduce biases related to breed, age, or image quality. Therefore, there is a need for more diverse and robust training samples to mitigate overfitting and enhance reliability in practical applications.

To address these limitations, future research will incorporate video data to capture dynamic movements and behavioral cues. It will help enable more accurate differentiation between KB-affected broilers and healthy ones through temporal analysis. Domain adaptation techniques will be explored to improve model transferability across varied housing systems, facilitating broader deployment in commercial settings. Moreover, integrating behavioral observations with structural assessments could enable early prediction of KB. In addition, associated with lameness issues, potentially through multi-modal approaches that correlate gait patterns, activity levels, and morphological features. These advancements aim to refine precision detection and support proactive welfare interventions in poultry production.

4. Conclusions

This study evaluated the effectiveness of various optimization parameters, preprocessing techniques, and YOLO model architectures for detecting KB. Among the tested optimizers, SGD achieved the highest precision and mAP scores, demonstrating superior localization and reduced false positives. Preprocessing comparisons revealed that using 960-pixel images provided the best trade-off between detection accuracy and computational cost. It achieved the highest F1 Score and the lowest FNR. Data augmentation also significantly improved recall and overall model robustness without increasing training time. Among YOLO architectures, YOLOv9 outperformed other YOLO models. Similarly, by optimizing and using augmented data with a 960-pixel image resolution, YOLOv9's performance was significantly improved. The optimized YOLOv9 model achieved near-perfect precision, recall, and F1-score. These results show how crucial it is to select the best optimizer, image resolution, augmentation, and fine-tune the model. Together, these choices enable highly accurate and reliable detection of poultry health conditions in real-world applications. Future research will utilize video data for enhanced KB detection and apply domain adaptation to diverse housing systems. It will also integrate behavioral and structural data to predict KB and lameness early, thereby improving accuracy and poultry welfare.

Author Contributions: R.B.B.: Writing—review and editing, Writing—original draft, Investigation, Visualization, Validation, Supervision, Software, Methodology, Formal analysis, Data curation, Conceptualization. Y.T. and C.P.: Writing—review and editing, Conceptualization. A.A., A.D.T.D. and A.A.K.A.: Writing—review and editing, Methodology, Data curation, Investigation. D.W.: Supervision, Writing—review and editing. All authors have read and agreed to the published version of the manuscript.

Funding: This research was supported and funded by the University of Arkansas Experimental Station and the University of Arkansas College of Engineering, USDA National Institute of Food and Agriculture (No: 2023-70442-39232, 2024-67022-42882).

Institutional Review Board Statement: Not applicable.

Informed Consent Statement: Not applicable.

Data Availability Statement: The original contributions presented in this study are included in the article. Further inquiries can be directed to the corresponding authors.

Acknowledgments: We acknowledge the computational support provided by the Arkansas High Performance Computing Center, which is made possible through multiple National Science Foundation grants and the Arkansas Economic Development Commission. We also used Grammarly to improve grammar and clarity throughout the manuscript.

Conflicts of Interest: The authors declare that no competing financial interests or personal relationships could have influenced the work presented in this paper.

References

1. Mottet, A.; Tempio, G. Global Poultry Production: Current State and Future Outlook and Challenges. *Worlds Poult. Sci. J.* **2017**, *73*, 245–256. [CrossRef]
2. USDA ERS. Poultry Expected to Continue Leading Global Meat Imports as Demand Rises | Economic Research Service. Available online: https://www.ers.usda.gov/amber-waves/2022/august/poultry-expected-to-continue-leading-global-meat-imports-as-demand-rises (accessed on 21 July 2025).
3. Tavárez, M.A.; Solis de los Santos, F. Impact of Genetics and Breeding on Broiler Production Performance: A Look into the Past, Present, and Future of the Industry. *Anim. Front.* **2016**, *6*, 37–41. [CrossRef]
4. Zuidhof, M.; Schneider, B.; Carney, V.; Korver, D.; Robinson, F. Growth, Efficiency, and Yield of Commercial Broilers from 1957, 1978, and 2005. *Poult. Sci.* **2014**, *93*, 2970–2982. [CrossRef] [PubMed]
5. Petracci, M.; Mudalal, S.; Soglia, F.; Cavani, C. Meat Quality in Fast-Growing Broiler Chickens. *Worlds Poult. Sci. J.* **2015**, *71*, 363–374. [CrossRef]
6. Che, S.; Wang, C.; Iverson, M.; Varga, C.; Barbut, S.; Bienzle, D.; Susta, L. Characteristics of Broiler Chicken Breast Myopathies (Spaghetti Meat, Woody Breast, White Striping) in Ontario, Canada. *Poult. Sci.* **2022**, *101*, 101747. [CrossRef]
7. Lilburn, M.S. Skeletal Growth of Commercial Poultry Species. *Poult. Sci.* **1994**, *73*, 897–903. [CrossRef]
8. Liu, K.; He, Y.; Xu, B.; Lin, L.; Chen, P.; Iqbal, M.K.; Mehmood, K.; Huang, S. Leg Disorders in Broiler Chickens: A Review of Current Knowledge. *Anim. Biotechnol.* **2023**, *34*, 5124–5138. [CrossRef]
9. Wise, D.R. Spondylolisthesis ('Kinky Back') in Broiler Chickens. *Res. Vet. Sci.* **1970**, *11*, 447–455. [CrossRef]
10. Anthney, A.; Do, A.D.T.; Alrubaye, A.A. Bacterial Chondronecrosis with Osteomyelitis Lameness in Broiler Chickens and Its Implications for Welfare, Meat Safety, and Quality: A Review. *Front. Physiol.* **2024**, *15*, 1452318. [CrossRef]
11. Choppa, V.S.R.; Kim, W.K. A Review on Pathophysiology, and Molecular Mechanisms of Bacterial Chondronecrosis and Osteomyelitis in Commercial Broilers. *Biomolecules* **2023**, *13*, 1032. [CrossRef]
12. Wideman, R.F., Jr. Bacterial Chondronecrosis with Osteomyelitis and Lameness in Broilers: A Review. *Poult. Sci.* **2016**, *95*, 325–344. [CrossRef] [PubMed]
13. Marzban Abbasabadi, B.; Golshahi, H.; Seifi, S. Pathomorphologial Investigation of Spondylolisthesis Leaded to Spondylosis in Commercial Broiler Chicken with Posterior Paralysis: A Case Study. *Vet. Res. Forum* **2021**, *12*, 263–265. [CrossRef] [PubMed]
14. Mitchell, R. *Broiler Lameness in the United States: An Industry Perspective*; Engormix: Córdoba, Argentina, 2014; Volume 175.
15. Riddell, C.; Howell, J. Spondylolisthesis ('Kinky Back') in Broiler Chickens in Western Canada. *Avian Dis.* **1972**, *16*, 444–452. [CrossRef] [PubMed]
16. Khan, M.A.; Olson, N.O.; Weiss, R. Spondylopathy in Broilers: Case Reports. *Poult. Sci.* **1973**, *52*, 1847–1852. [CrossRef]
17. Dinev, I. Pathomorphological Investigations on the Incidence of Clinical Spondylolisthesis (Kinky Back) in Different Commercial Broiler Strains. *Rev. Med. Vet.* **2012**, *163*, 511–515.
18. Pompeu, M.; Barbosa, V.; Martins, N.; Baião, N.; Lara, L.; Rocha, J.; Miranda, D. Nutritional Aspects Related to Non-Infectious Diseases in Locomotor System of Broilers. *Worlds Poult. Sci. J.* **2012**, *68*, 669–678. [CrossRef]
19. Lourenço da Silva, M.I.; Leonie, J. *Kinky Back (Spondylolisthesis) in Broiler Chickens: What We Can Do Today to Reduce the Problem*; VCE Publications: Blacksburg, VA, USA, 2023.
20. Robbins, K.M.; Suyemoto, M.M.; Lyman, R.L.; Martin, M.P.; Barnes, H.J.; Borst, L.B. An Outbreak and Source Investigation of Enterococcal Spondylitis in Broilers Caused by *Enterococcus cecorum*. *Avian Dis.* **2012**, *56*, 768–773. [CrossRef]
21. Borst, L.; Suyemoto, M.; Sarsour, A.; Harris, M.; Martin, M.; Strickland, J.; Oviedo, E.; Barnes, H. Pathogenesis of Enterococcal Spondylitis Caused by *Enterococcus cecorum* in Broiler Chickens. *Vet. Pathol.* **2017**, *54*, 61–73. [CrossRef]
22. Lourenço da Silva, M.I.; de Lima Almeida Paz, I.C.; Jacinto, A.S.; Nascimento Filho, M.A.; de Oliveira, A.B.S.; dos Santos, I.G.A.; dos Santos Mota, F.; Caldara, F.R.; Jacobs, L. Providing Environmental Enrichments Can Reduce Subclinical Spondylolisthesis Prevalence without Affecting Performance in Broiler Chickens. *PLoS ONE* **2023**, *18*, e0284087. [CrossRef]
23. Anthney, A. Bacterial Chondronecrosis with Osteomyelitis in Broiler Chickens: Experimental Lameness and Additive Probiotic Treatments. Master's Thesis, University of Arkansas, Fayetteville, AR, USA, 2025.
24. Bist, R.B.; Yang, X.; Subedi, S.; Bist, K.; Paneru, B.; Li, G.; Chai, L. An Automatic Method for Scoring Poultry Footpad Dermatitis with Deep Learning and Thermal Imaging. *Comput. Electron. Agric.* **2024**, *226*, 109481. [CrossRef]
25. Derakhshani, S.M.; Overduin, M.; van Niekerk, T.G.; Groot Koerkamp, P.W. Implementation of Inertia Sensor and Machine Learning Technologies for Analyzing the Behavior of Individual Laying Hens. *Animals* **2022**, *12*, 536. [CrossRef]
26. Manikandan, V.; Neethirajan, S. Decoding Poultry Welfare from Sound—A Machine Learning Framework for Non-Invasive Acoustic Monitoring. *Sensors* **2025**, *25*, 2912. [CrossRef]
27. Subedi, S.; Bist, R.B.; Yang, X.; Li, G.; Chai, L. Advanced Deep Learning Methods for Multiple Behavior Classification of Cage-Free Laying Hens. *AgriEngineering* **2025**, *7*, 24. [CrossRef]
28. Dayan, J.; Goldman, N.; Waiger, D.; Melkman-Zehavi, T.; Halevy, O.; Uni, Z. A Deep Learning-Based Automated Image Analysis for Histological Evaluation of Broiler Pectoral Muscle. *Poult. Sci.* **2023**, *102*, 102792. [CrossRef] [PubMed]

29. Marzoque, H.J.; de Alencar Nääs, I.; Marzoque, G.A.; do Carmo, M.; de Alencar, B.; Batista, M.L. Design of an Application for Analysing the Progression of Musculoskeletal Diseases in Slaughterhouses. In Proceedings of the XL CIOSTA & CIGR Section V International Conference, Évora, Portugal, 10–13 September 2023; Volume 158.
30. Brossard, L.; Quiniou, N.; Marcon, M.; Méda, B.; Dusart, L.; Lopez, V.; Dourmad, J.-Y.; Pomar, J. *Development of a Decision Support System for Precision Feeding Application in Pigs and Poultry*; HAL: Lyon, France, 2017; Volume 2017.
31. Kader, M.S.; Ahmed, F.; Akter, J. Machine Learning Techniques to Precaution of Emerging Disease in the Poultry Industry. In Proceedings of the 2021 24th International Conference on Computer and Information Technology (ICCIT), Dhaka, Bangladesh, 18–20 December 2021; IEEE: Dhaka, Bangladesh, 2021; pp. 1–6.
32. Yang, X.; Bist, R.B.; Paneru, B.; Chai, L. Deep Learning Methods for Tracking the Locomotion of Individual Chickens. *Animals* **2024**, *14*, 911. [CrossRef] [PubMed]
33. Pallerla, C.; Feng, Y.; Owens, C.M.; Bist, R.B.; Mahmoudi, S.; Sohrabipour, P.; Davar, A.; Wang, D. Neural Network Architecture Search Enabled Wide-Deep Learning (NAS-WD) for Spatially Heterogenous Property Awared Chicken Woody Breast Classification and Hardness Regression. *Artif. Intell. Agric.* **2024**, *14*, 73–85. [CrossRef]
34. Nasiri, A.; Yoder, J.; Zhao, Y.; Hawkins, S.; Prado, M.; Gan, H. Pose Estimation-Based Lameness Recognition in Broiler Using CNN-LSTM Network. *Comput. Electron. Agric.* **2022**, *197*, 106931. [CrossRef]
35. Bao, Y.; Lu, H.; Zhao, Q.; Yang, Z.; Xu, W. Detection System of Dead and Sick Chickens in Large Scale Farms Based on Artificial Intelligence. *Math. Biosci. Eng.* **2021**, *18*, 6117–6135. [CrossRef]
36. Bist, R.B.; Subedi, S.; Yang, X.; Chai, L. Automatic Detection of Cage-Free Dead Hens with Deep Learning Methods. *AgriEngineering* **2023**, *5*, 1020–1038. [CrossRef]
37. Yang, X.; Zhao, Y.; Gan, H.; Hawkins, S.; Eckelkamp, L.; Prado, M.; Burns, R.; Purswell, J.; Tabler, T. Modeling Gait Score of Broiler Chicken via Production and Behavioral Data. *Animal* **2023**, *17*, 100692. [CrossRef]
38. Jaihuni, M.; Zhao, Y.; Gan, H.; Tabler, T.; Qi, H. Automated Broiler Mobility Evaluation Through DL and ML Models: An Alternative Approach to Manual Gait Assessment. *AgriEngineering* **2025**, *7*, 133. [CrossRef]
39. Makrai, L.; Nemes, C.; Simon, A.; Ivanics, E.; Dudás, Z.; Fodor, L.; Glávits, R. Association of *Enterococcus cecorum* with Vertebral Osteomyelitis and Spondylolisthesis in Broiler Parent Chicks. *Acta Vet. Hung.* **2011**, *59*, 11–21. [CrossRef]
40. Asnayanti, A.; Do, A.D.T.; Alharbi, K.; Alrubaye, A. Inducing Experimental Bacterial Chondronecrosis with Osteomyelitis Lameness in Broiler Chickens Using Aerosol Transmission Model. *Poult. Sci.* **2024**, *103*, 103460. [CrossRef] [PubMed]
41. Hassan, E.; Shams, M.Y.; Hikal, N.A.; Elmougy, S. The Effect of Choosing Optimizer Algorithms to Improve Computer Vision Tasks: A Comparative Study. *Multimed. Tools Appl.* **2023**, *82*, 16591–16633. [CrossRef] [PubMed]
42. Tong, Q.; Liang, G.; Bi, J. Calibrating the Adaptive Learning Rate to Improve Convergence of ADAM. *Neurocomputing* **2022**, *481*, 333–356. [CrossRef] [PubMed]
43. Zhou, P.; Xie, X.; Lin, Z.; Yan, S. Towards Understanding Convergence and Generalization of AdamW. *IEEE Trans. Pattern Anal. Mach. Intell.* **2024**, *46*, 6486–6493. [CrossRef]
44. Amir, I.; Koren, T.; Livni, R. SGD Generalizes Better than GD (and Regularization Doesn't Help). *Proc. Mach. Learn. Res.* **2021**, *134*, 63–92.
45. Gupta, A.; Ramanath, R.; Shi, J.; Keerthi, S.S. *Adam vs. Sgd: Closing the Generalization Gap on Image Classification*; LinkedIn: Sunnyvale, CA, USA, 2021; pp. 1–7.
46. Rebuffi, S.-A.; Gowal, S.; Calian, D.A.; Stimberg, F.; Wiles, O.; Mann, T.A. Data Augmentation Can Improve Robustness. *Adv. Neural Inf. Process. Syst.* **2021**, *34*, 29935–29948.
47. Bist, R.B.; Subedi, S.; Yang, X.; Chai, L. A Novel YOLOv6 Object Detector for Monitoring Piling Behavior of Cage-Free Laying Hens. *AgriEngineering* **2023**, *5*, 905–923. [CrossRef]
48. Ultralytics Models Supported by Ultralytics. Available online: https://docs.ultralytics.com/models (accessed on 28 December 2024).
49. Liu, D.; Wang, B.; Peng, L.; Wang, H.; Wang, Y.; Pan, Y. HSDNet: A Poultry Farming Model Based on Few-Shot Semantic Segmentation Addressing Non-Smooth and Unbalanced Convergence. *PeerJ Comput. Sci.* **2024**, *10*, e2080. [CrossRef]
50. Zhou, P.; Feng, J.; Ma, C.; Xiong, C.; Hoi, S.; E, W. Towards Theoretically Understanding Why SGD Generalizes Better Than Adam in Deep Learning. *arXiv* **2020**, arXiv:2010.05627. [CrossRef]
51. Subedi, S.; Bist, R.; Yang, X.; Chai, L. Tracking Pecking Behaviors and Damages of Cage-Free Laying Hens with Machine Vision Technologies. *Comput. Electron. Agric.* **2023**, *204*, 107545.
52. Saponara, S.; Elhanashi, A. Impact of Image Resizing on Deep Learning Detectors for Training Time and Model Performance. In Proceedings of the Applications in Electronics Pervading Industry, Environment and Society; Saponara, S., De Gloria, A., Eds.; Springer International Publishing: Cham, Switzerland, 2022; pp. 10–17.

53. Kumar, T.; Brennan, R.; Mileo, A.; Bendechache, M. Image Data Augmentation Approaches: A Comprehensive Survey and Future Directions. *IEEE Access* **2024**, *12*, 187536–187571. [CrossRef]
54. Poojary, R.; Raina, R.; Kumar Mondal, A. Effect of Data-Augmentation on Fine-Tuned CNN Model Performance. *IAES Int. J. Artif. Intell.* **2021**, *10*, 84. [CrossRef]

AgriEngineering

MDPI

Article

A Machine Vision System for Monitoring Wild Birds on Poultry Farms to Prevent Avian Influenza

Xiao Yang [1], Ramesh Bahadur Bist [1], Sachin Subedi [1], Zihao Wu [2], Tianming Liu [2], Bidur Paneru [1] and Lilong Chai [1,*]

[1] Department of Poultry Science, College of Agricultural & Environmental Sciences, University of Georgia, Athens, GA 30602, USA

[2] Department of Computer Science, Franklin College of Arts and Sciences, University of Georgia, Athens, GA 30602, USA

* Correspondence: lchai@uga.edu

Abstract: The epidemic of avian influenza outbreaks, especially high-pathogenicity avian influenza (HPAI), which causes respiratory disease and death, is a disaster in poultry. The outbreak of HPAI in 2014–2015 caused the loss of 60 million chickens and turkeys. The most recent HPAI outbreak, ongoing since 2021, has led to the loss of over 50 million chickens so far in the US and Canada. Farm biosecurity management practices have been used to prevent the spread of the virus. However, existing practices related to controlling the transmission of the virus through wild birds, especially waterfowl, are limited. For instance, ducks were considered hosts of avian influenza viruses in many past outbreaks. The objectives of this study were to develop a machine vision framework for tracking wild birds and test the performance of deep learning models in the detection of wild birds on poultry farms. A deep learning framework based on computer vision was designed and applied to the monitoring of wild birds. A night vision camera was used to collect data on wild bird near poultry farms. In the data, there were two main wild birds: the gadwall and brown thrasher. More than 6000 pictures were extracted through random video selection and applied in the training and testing processes. An overall precision of 0.95 (mAP@0.5) was reached by the model. The model is capable of automatic and real-time detection of wild birds. Missed detection mainly came from occlusion because the wild birds tended to hide in grass. Future research could be focused on applying the model to alert to the risk of wild birds and combining it with unmanned aerial vehicles to drive out detected wild birds.

Keywords: wildlife monitoring; biosecurity; avian influenza; bird detection; deep learning

1. Introduction

High-pathogenicity avian influenza (HPAI) is disastrous for poultry industries worldwide. The outbreak of HPAI in 2014–2015 caused the loss of 60 million chickens and turkey [1]. The most recent HPAI outbreak, ongoing since 2021, has led to the loss of over 50 million chickens so far in the US and Canada [2]. Farm biosecurity management practices have been applied to prevent the spread of the virus [3]. Wild birds, especially some waterfowl and shorebirds, are the natural hosts of avian influenza (AI) virus. They can be infected with HPAI and show no symptoms. However, their droppings spread AI virus to domestic flocks during migration [4,5]. Avian influenza, also known as bird flu, is a viral infection that affects bird species, particularly wild aquatic birds and domestic poultry like chickens. It is categorized into low-pathogenic avian influenza (LPAI) and highly pathogenic avian influenza (HPAI) based on the severity of the disease they cause [6]. LPAI generally results in no noticeable symptoms or only mild ones in chickens, whereas HPAI leads to severe disease with high mortality rates. Of significant concern are the H5 and H7 strains of LPAI, which have the potential to mutate into the deadly HPAI. This mutation can occur when the virus circulates within a poultry population, undergoing changes to its hemagglutinin protein, which then increase its virulence [7,8]. HPAI has a higher mortality rate in chickens than LPAI and it spreads rapidly; it could even be passed to humans [9].

Wild birds are attracted by feeds, shelters, and dams on poultry farms. However, monitoring and preventing the AI risk carried by wild birds is challenging at present because of inadequate AI outbreak data and the uncertain movements of nomadic and local wild birds, which are sensitive to weather changes [10]. Poultry farms can inadvertently become hotspots for the spread of AI as they attract wild birds seeking food. These wild birds, natural reservoirs for AI viruses, can carry and transmit the virus to domesticated birds through direct or indirect contact. For example, in the United States, during the 2022–2023 outbreak, more than 8500 detections of highly pathogenic avian influenza (HPAI) were recorded in wild birds, contributing to over 1000 confirmed outbreaks in commercial and backyard flocks. These outbreaks resulted in the loss of approximately 77.9 million birds across 47 states [11]. The primary concern is the contamination of chicken feed and water through the feces of infected wild birds. Such contamination can lead to rapid transmission of the virus among the farm's poultry population. Due to the close quarters and high density of commercial chicken farms, the disease can spread swiftly, with the potential for significant economic losses, as well as public health concerns. Therefore, strict biosecurity measures, including the control of wild birds' access to poultry feeding and watering areas, are crucial to prevent the introduction and spread of AI [12,13]. Humans, wild birds, and poultry have shown cross-species transmission of AI viruses in poultry houses or nearby areas (Figure 1).

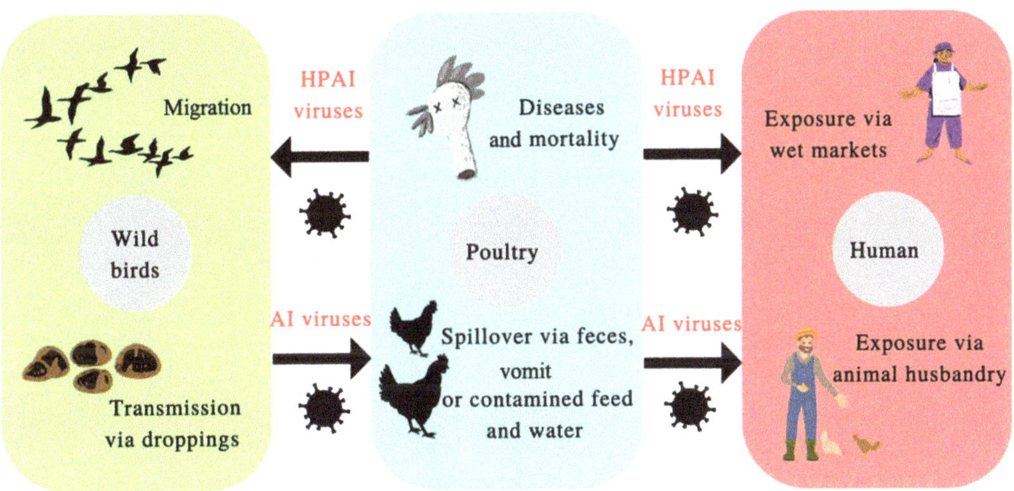

Figure 1. Generalized transmission of avian influenza viruses.

The automated detection of wild birds provides an effective way to understand AI transmission pathways [14,15]. The detection system includes sensors (cameras, robots, or radar) to collect target data, training (feature extraction and selection), and classification. Radar-based detection systems have been developed to detect birds for many years. However, their limitations are as follows: (a) identifying birds is impossible; (b) some specific radiofrequencies and microwaves can cause electromagnetic interference with other electronic equipment in poultry houses [16,17]. Therefore, in this study, we investigated an emerging computer vision system to monitor and classify wild birds. Compared with traditional detection systems, computer vision improves the overall performance of the system by introducing deep learning methods. You only look once (YOLO) is one of the most popular deep learning methods, famous for its high precision and real-time detection [18]. Datar et al. [19] developed a YOLOv3 model to detect wild birds in the wild which reached an accuracy of 0.87 on average [19]. Human observation costs are generally limited to salaries, typically around USD 15–25 per hour, while telescopes can range from USD 100–1000 depending on quality. In contrast, setting up a computer vision system involves higher upfront costs. Cameras and graphics processing units (GPUs) can cost between USD 2000–15,000, and additional software or model development may add further expenses. However, automatic systems can save on labor costs and make automatic deterrents possible [20].

However, most researchers have only focused on applying YOLO to monitor wild birds, and the YOLO they utilized has since been updated. When it comes to poultry farms, the situation is dramatically different. Currently, the egg industry is shifting to cage-free houses to improve chicken welfare [3]. Under cage-free situations, hens have more chances to behave naturally and come into contact with wild birds, causing AI transmission [21–25]. Therefore, in our study, we used YOLOv5 and YOLOX, which are the newest YOLO series models to date. In addition, we built a computer vision-based method to track wild birds near cage-free houses to prevent HPAI transmission.

The objectives of this study were as follows: (1) to develop detectors for tracking wild birds near cage-free houses using the updated YOLO methods; (2) to compare the model performance between YOLOX and YOLOv5; (3) to predict and manage the presence of wild birds to improve flock health based on the method developed in this study.

2. Materials and Methods

2.1. Experimental Design and Data Collection

A machine vision system using night vision cameras (Swann Communications USA Inc., Santa Fe Springs, CA, USA) and deep learning models was developed in this study. The system includes a 1080p full high-definition camera (PRO-1080MSFB) with a focal length ranging from 2.8 mm to 3.6 mm capturing 18 frames per second (FPS) with a resolution of 1440 pixels high and 1080 pixels wide. The installation height of the camera was 1.5 m. The camera used a digital video recorder (DVR-5580) and a computer (Dell-P2419H) and was located at the poultry research center M house at the University of Georgia (UGA), Athens, GA, USA. The system monitored wild birds automatically 24 h a day. Utilizing night vision in low-light conditions helped us record wild bird activity during the evening and early morning. Night vision cameras are essential for capturing data in low-light environments where birds may be active and where traditional cameras would struggle to provide clear footage. The videos are then saved to the recorder (Figure 2). In the Department of Poultry Science at UGA, the protocol for managing video data entails a systematic transfer from recording devices to hard storage, centralizing data for analysis. This process was meticulously overseen by team members who conducted daily inspections of both the camera's position and the lens's condition to ensure consistent quality throughout the experimental phase. Spanning from 14 August to 10 September 2022, this month-long data collection phase was critical for capturing the nuances of poultry behavior. Following this period, a random selection of the video data was converted into static images using the Free Video Converter software (1.6.1). This conversion marked a preparatory step for the next phase of this study, allowing for detailed and focused analysis of the imagery. The converted images are instrumental for researchers, serving as a foundation for further examination and insights into poultry science, contributing to advancements in the field.

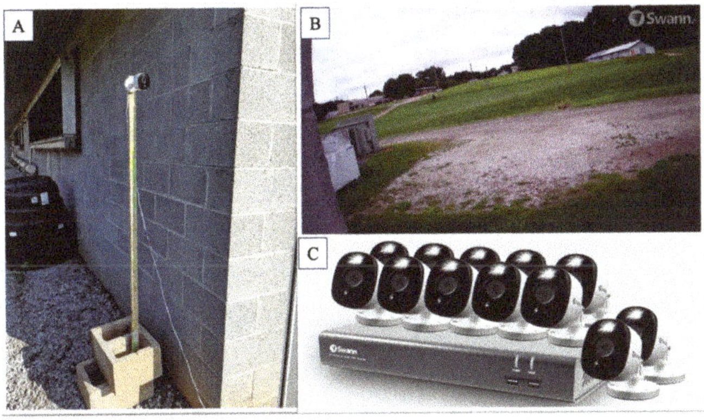

Figure 2. Swann monitor system for collecting wild birds' images/videos. Camera setup outside a poultry house for monitoring wild birds (**A**). Sample footage captured from the Swann camera showing the monitoring area (**B**). The Swann camera system used for the experiment, including multiple cameras and the recording unit (**C**).

2.2. *Dataset Augmentation Strategy*

To bolster the efficacy and training performance of machine learning models, four fundamental augmentation techniques are used. First, scaling adjusts the size of images to a specified dimension, ensuring consistency across the dataset. Cropping is utilized to extract and focus on specific parts of an image, which can aid in the model's ability to recognize features from various angles and sections. Flipping is another technique used, where images are mirrored horizontally, providing the model with a wider variety of orientations to learn from. Lastly, rotation introduces a range of angular perspectives by turning the images by random degrees. Each method contributes to a more robust training set, simulating a diverse array of possible scenarios and appearances that the model may encounter in real-world applications. This diversity is crucial for developing an AI system capable of an accurate and reliable detection of wild birds [26,27]. After image augmentation, a total number of 6612 images were gained, of which 5290 images were divided into testing dataset. Then, 1322 images were used as a valid dataset [28,29]. The details of dataset augmentation were shown below (Figure 3). The augmented images were labeled with Labeling Window v2.1.1. The wild birds were labeled with a bounding box in each picture, which contained at least 1/3 their body size. The wild bird class name was given after labeling.

Figure 3. The process of dataset augmentation, showing various transformations applied to the original image to enhance the training dataset. The augmentations include scaling, cropping, flipping, and rotation.

2.3. Detection of Wild Birds

In the field of computer vision for wildlife monitoring, this study harnesses the power of two advanced YOLO algorithms, YOLOv5 and YOLOX, to detect wild birds with enhanced precision. YOLOv5 employs an anchor-based bounding box methodology to pinpoint the location of objects within an image, leveraging predefined box dimensions as references to aid in object detection. This approach, while effective, contrasts with the innovative strategy adopted by YOLOX, which eschews anchor boxes entirely. YOLOX's anchor-free framework achieves end-to-end detection by utilizing a key point head that identifies central points of objects for localization. This key point-based strategy closely mirrors human visual perception, where the eye naturally seeks out notable points to discern objects in the visual field. The anchor-free detector's methodology resonates with this intuitive process, initially establishing key points that act as focal indicators to outline the object's area. Subsequently, the algorithm applies attention mechanisms to refine the detection, concentrating on the central region to optimize accuracy. This center-based approach enables YOLOX to not only maintain but also improve upon the accuracy levels of traditional YOLO series feature pyramids. By adopting this more human-like perception model, YOLOX pushes the boundaries of object detection, offering a sophisticated, refined tool for the precise identification and tracking of avian species in their natural habitats, a critical advancement for ecological research and the conservation of biodiversity [30–33].

The YOLOX architecture offers a suite of seven different models designed to accommodate various computational and resource requirements, each tailored to specific use cases. The five standard structures—YOLOX-s, YOLOX-m, YOLOX-l, YOLOX-x, and YOLOX-Darknet53—provide a range of options from small to extra-large sizes, catering to diverse performance needs. Each variant is distinguished by its input image size, the number of parameters defining the model's complexity, and the number of floating-point operations per second (FLOPs) it requires (Table 1). The larger models, like YOLOX-l and YOLOX-x, offer higher accuracy due to more parameters and increased FLOPs, making them suitable for systems with ample computational power. Conversely, YOLOX-s and YOLOX-m are designed for environments where efficiency is prioritized over high precision. Complementing these are the two light models, YOLOX-Nano and YOLOX-Tiny, engineered for scenarios demanding ultra-fast inference times and minimal resource usage, such as mobile or embedded devices. These models significantly reduce the computational load by decreasing the input image size and parameters, resulting in fewer FLOPs. The trade-off, however, is a reduction in detection accuracy compared to their standard counterparts. These light models are particularly advantageous in real-time applications where speed is critical and computational resources are limited. Collectively, the YOLOX suite demonstrates the versatility of the YOLO family, offering solutions that range from high-accuracy, resource-intensive models to lightweight, speed-optimized versions, thus addressing the varied demands of modern object detection tasks [34,35]. The YOLOX network structure includes three main parts, which are the backbone, neck, and head (Figure 4). The backbone, using the same Darknet53 as the YOLOv3 algorithm and the feature pyramid network (FPN), is used to extract information from an image with different aspects. Essentially, the Darknet53 backbone outputs three feature layers (256 channels, 512 channels and 1024 channels) into the neck part [36,37]. The YOLOX neck updates a coupled head (original YOLO) to a decoupled head and outputs three tensors for better average precision. The three tensor outputs aggregate the class name of the bounding box (Cls), the dimension of the bounding box (X: x coordinate of the center of the bounding box, Y: y coordinate of the center of the bounding box, W: width of the bounding box and H: heigh of the bounding box), and the intersection over union (IoU) for prediction [38]. In the prediction part, different H $\times$ W predictions were used to predict whether there was an object or not at the center point in each layer, the class name of the object, and the position of the object.

Table 1. The differences between YOLOX models.

Model	Size	Parameters (M)	FLOPs (G)
YOLOX-s	640	9.0	26.8
YOLOX-m	640	25.3	73.8
YOLOX-l	640	54.2	155.6
YOLOX-x	640	99.1	281.9
YOLOX-Darknet53	640	63.7	185.3
YOLOX-Nano	416	0.91	1.08
YOLOX-Tiny	416	5.06	6.45

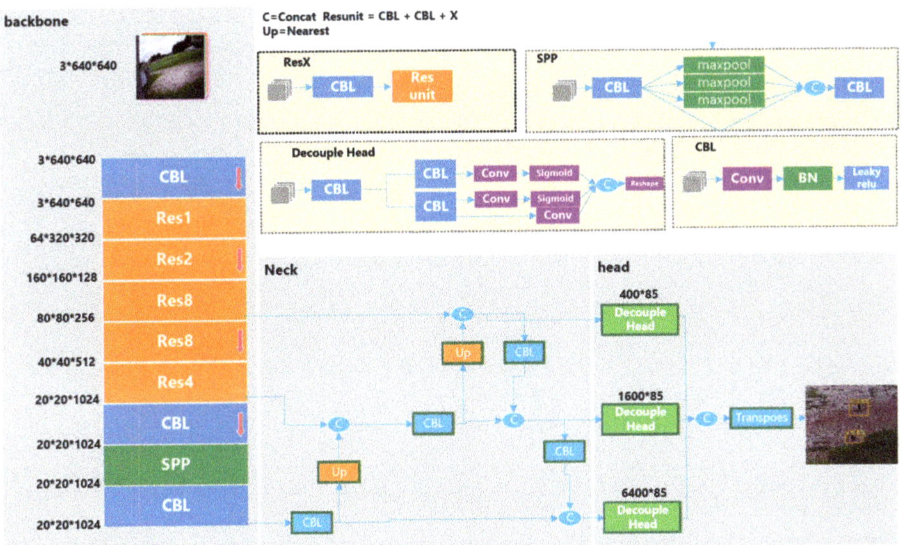

Figure 4. The structure of YOLOX.

The YOLOX object detection model represents a cutting-edge integration of machine learning into image recognition tasks, and its training was executed within the versatile and accessible confines of Google Colab. This cloud-based environment, built upon the Jupyter notebook interface, allows for the seamless execution of complex computational tasks without the need for local resource-intensive setups. Leveraging the cloud's computational power, the YOLOX model was trained using NVIDIA's CUDA toolkit, version 11.2, a parallel computing platform and programming model that significantly enhances processing efficiency. Key to the system's performance is the NVIDIA-SMI graphics card with 16 GB of memory, which is particularly adept at handling the parallel processing demands of machine learning algorithms. This robust hardware enables the handling of large volumes of data with remarkable speed, a crucial factor given the extensive nature of the training data involved. To optimize the training process, YOLOX was pretrained on the COCO dataset, a large-scale object detection, segmentation, and captioning dataset. This pretraining imbued the model with a foundational understanding of various object types and features, thereby reducing the overall time required for training the model to detect specific objects, such as wild birds. The model's training regimen was configured with a batch size of 8, which dictates the number of training samples to be processed before the model's internal parameters are updated. The choice of batch size strikes a balance between memory usage and the granularity of the update process. With 300 epochs set for the training duration, the model was exposed to the dataset multiple times, allowing it to iteratively learn and refine its detection capabilities through repeated passes. This extensive training ensured that the model reached a high level of accuracy, learning from the vast variety of images and annotations contained within the COCO dataset. Such a sophisticated training procedure allowed YOLOX to achieve a state of readiness where it can accurately identify and classify objects within images, a capability that is particularly beneficial for applications that require the rapid and reliable detection of objects in diverse and challenging environments. The use of Google Colab and the advanced hardware and software specifications detail a training environment setup for rigorous, extensive, and deep learning tasks, reflecting the progressive trajectory of machine learning and its applications in the real world.

2.4. Evaluation Standard for Monitoring Wild Birds

To evaluate the performance of wild bird detectors between YOLOX and YOLOv5, several standard indicators were set next, including true positives (TPs), false negatives (FNs), false positives (FPs) and true negatives (TNs) [39]. Precision, recall, and F1 score are critical indicators of a model's performance. Precision represents the accuracy of the model in classifying an image as positive, signifying the fraction of true positive identifications out of all positive identifications made. Recall, on the other hand, measures the model's ability to correctly identify all available actual positives, essentially quantifying the sensitivity of the model to detecting positives. The F1 score emerges as a balance between precision and recall, providing a single score that weighs both the model's precision and recall equally through their harmonic mean; this is particularly useful when seeking a balance between the model's performance in precision and sensitivity. Beyond these, average precision (AP) and mean average precision (mAP) extend the evaluation to a broader scope. AP calculates the accuracy of the detector for a single class across all recall levels, integrating the precision–recall curve. The mAP then takes this a step further by averaging the AP across all classes, offering an overarching measure of the detector's accuracy across different bird species [25]. By comparing these metrics, we can determine which bird detector performs better.

$$Precision = \frac{TPs}{(TPs + FPs)} \tag{1}$$

$$Recall = \frac{TPs}{(TPs + FNs)} \tag{2}$$

$$F1 = \frac{2 * Precision * Recall}{(Precision + Recall)} \tag{3}$$

$$AP = \int_{r=0}^{1} P(r)dr \tag{4}$$

where average precision is the area of the *Precision* and *Recall* curve

$$mAP = \frac{\sum_{i=1}^{N} AP_i}{N} \tag{5}$$

where N is the sum number of classes in the dataset.

To quantify the movement of activity patterns in our study, we used the YOLO model to detect species from video footage. The detection results were exported as a CSV file, which was then imported into Excel. In Excel, we calculated the average detections per hour for each species and created a line chart to visualize the activity patterns. This chart illustrated the movement trends of species throughout the day, allowing us to quantify and compare their activity levels over time.

3. Results and Discussions

3.1. The Results and Comparisons of Different Algorithms

To attest the superiority of the YOLOX, YOLOv5 was used as a representation of the previous YOLO series model for comparative study. All models were trained and tested with same dataset to keep uniformity. The consequences of the comparison experiments are shown below (Figure 5). The mAP@0.5 (when IoU threshold is 0.5) values of the two models were close before 250 epochs. However, YOLOX markedly exceeded YOLOv5 after 250 epochs with the highest mAP (95.4%), which improved the original YOLOv5 (86.4%) by 9%. To better comprehend the comparative results, three different losses were summarized to evaluate the model performance. The loss functions are formed of three metrics, box_loss (bounding box regression loss), obj_loss (the confidence of object presence), and cls_loss (classification loss). The box_loss of YOLOX was continually lower than that of YOLOv5, which indicates YOLOX outputs more correctly than YOLOv5 because box_loss measures how wrong the model is in evaluating the relationship between x and y. YOLOX applies the decoupled head to capture more channel features [40]. In terms of obj_loss, the YOLOX and YOLOv5 algorithms shared the same trend and featured very close loss values before 250 epochs, but the YOLOX method was slightly improved between 250 to 300 epochs, which means its confidence in predicting that the object exists in a proposed region of interest was higher. For cls_loss, the loss value of both methods was constant. The reason for this is that the dataset has only one class, so there is no missed class. YOLOX can improve the overall accuracy and confidence in detection.

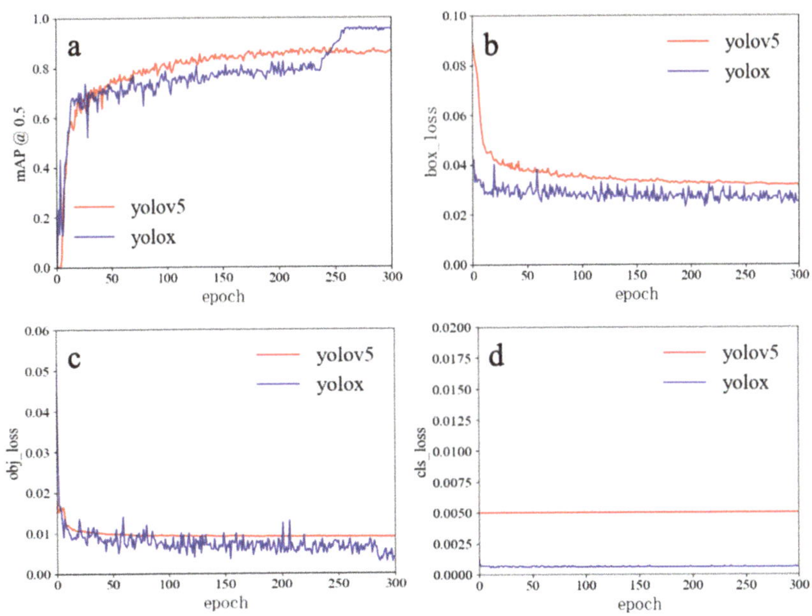

Figure 5. The results of comparative experiments. ((**a**): mAP @ 0.5 (Mean Average Precision) vs. Epoch, (**b**): Box Loss vs. Epoch, (**c**): Objective Loss vs. Epoch, (**d**): Classification Loss vs. Epoch).

Two detection examples based on the two different models are shown in Figures 6 and 7. Both models can detect and locate gadwalls (a common reservoir of H5N1 virus) and brown thrashers (a potential host of HPAI virus) correctly [41,42]. However, the detection confidence of YOLOX is higher than that of YOLOv5 (on average, it is 10% higher), which also indirectly reflects the fact that YOLOX has better accuracy. The main advantage of YOLOX over YOLOv5 lies in its architectural improvements. YOLOX incorporates an anchor-free algorithm, which simplifies the detection process by eliminating the need for predefined anchor boxes. This leads to a more efficient detection of objects, especially those of varying sizes and shapes. Additionally, YOLOX leverages advanced detection techniques such as a decoupled head and simple optimal transport assignment (SimOTA), which further enhance its performance. Nevertheless, the models also generated some flaws when the bodies of wild birds were hidden in the grass on the ground or by other wild birds. This is a common error in object detection, especially in open-field areas, because there are a number of shelters and natural objects overlapping with the target objects [43,44]. In addition, object size also affects detection accuracy. Wild birds, especially the brown thrasher, are small birds. To improve model performance in small bird detection, high-resolution images could be applied in the prediction process [45]. Alternatively, other solutions such as utilizing additional camera angles or deploying unmanned aerial vehicles (UAVs) could be explored to address this challenge. In this study, other errors included humans interfering with wild birds, which drove the wild birds away from the place we observed them in. However, this could not be prevented because farm workers and researchers conducting experiments here have free access to their experimental houses. In addition, model performance is affected by weather conditions. Rainfall makes observing wild birds difficult because wild birds return to their shelters. However, even in sunny weather, there are also some limitations. The sun's light causes shadows, and when shadows cover wild birds, this makes the detected targets blurry [46]. To solve this issue, a dual-cue network is proposed to recover shadows in local pictures [47,48]. Region-enhanced networks with normalized attention are an alternative way to address the issue of shadows [49]. In addition, deep metric leaning combined with knowledge graphs would provide an efficient way to detect similar objects [50–52]. Although there are still some flaws in the automated detection system we used, its cornerstone (model accuracy) achieved acceptable results.

Figure 6. Automatic detection of the gadwall by our machine vision model. The left-hand image shows a gadwall. The top-right image displays YOLOX's detection with bounding boxes and confidence scores. The bottom-right image shows YOLOv5's detection for comparison.

Figure 7. Automatic detection of the brown thrasher by our machine vision model. The left-hand image shows a reference brown thrasher. The top-right image displays YOLOX's detection, with bounding boxes and confidence scores. The bottom-right image shows YOLOv5's detection for comparison.

3.2. Identification and Classification of Wild Birds

In the study of wildlife and agricultural coexistence, the behavior of wild birds is a critical factor, particularly in relation to the transmission of diseases such as highly pathogenic avian influenza [53]. This comprehensive study zeroes in on two avian species, the gadwall and the brown thrasher, which were meticulously tracked and classified by a sophisticated detection model (Figure 8).

Figure 8. Detection of brown thrashers and gadwalls in evening conditions using the yolo model.

The gadwall is a modest duck, often overshadowed by more colorful fowl, yet it plays a significant role in disease ecology. These ducks frequent wetlands, the very environments that are prime for the spread of HPAI, mingling in the shared waters that can serve as conduits for the virus. The gadwall's migration patterns and feeding habits—dabbling in surface waters where the virus may be present—position it as a potential vector for AI, bridging wild ecosystems and domestic settings [54].

On the other hand, the brown thrasher, distinguished by its streaked feathers and melodious calls, operates closer to the ground. Thriving in dense brushwood, the thrasher's foraging can disturb the soil and unearth the AI virus, potentially held within organic debris. Its terrestrial habits bring it into closer contact with poultry farms, thus implicating it in the vector chain [55].

From 6 to 8 September, snapshots of these birds' activities were captured every hour. These snapshots highlight typical behavioral patterns that can be identified throughout the entire data collection period. The gadwall showed greater numbers at most times of the day, except for a noticeable dip in the early afternoon—perhaps due to a siesta or a shift to a feeding ground less visible to the cameras. The brown thrasher, however, peaked in the early afternoon, suggesting a period of heightened activity that could correlate with increased AI transmission risk (Figure 9). In this study, weather conditions did not significantly impact the observation of wild birds. During periods of heavy rain or fog, wild birds tended to remain in their habitats and were less likely to emerge.

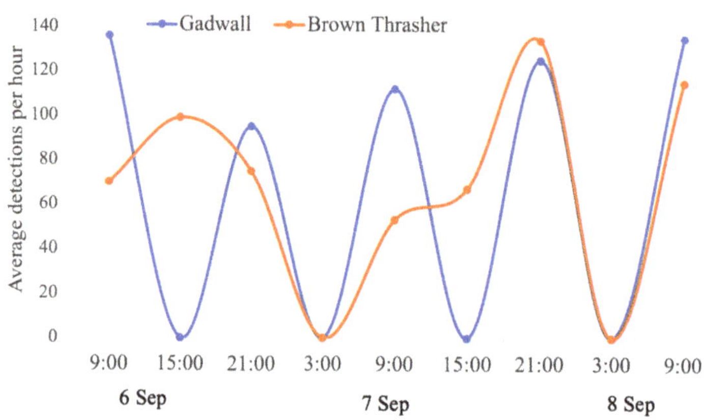

Figure 9. Activity of wild birds per hour detected by machine vision model.

The findings extend beyond the academic sphere, carrying weighty implications for farm management and disease prevention. Recognizing when wild birds are most active can inform farming practices to mitigate HPAI risks, like adjusting poultry's outdoor access or reinforcing biosecurity measures during peak bird activity times. This data-driven approach underscores the importance of predictive behavioral modeling in managing disease vectors. By leveraging technology, we can obtain a deeper understanding of wild bird behavior, enhancing our ability to coexist sustainably with wildlife while protecting domestic bird populations from diseases like HPAI.

3.3. Strategies to Prevent Avian Influenza

In the bid to mitigate the risk of avian influenza transmission from wild birds to poultry, integrating deterrent strategies with advanced detection models like YOLOX provides a sophisticated approach to wildlife management on farms. The proposed deterrence cycle commences with the target phase, where the YOLOX model efficiently detects wild birds, pinpointing species known as AI hosts, such as the gadwall and brown thrasher. The model's precision in identifying these specific species allows for a focused and informed response. Following detection, the track phase involves the model summarizing data to locate areas of high bird activity. By mapping these frequent visitation spots, the model provides valuable insights, allowing farmers to understand the movement patterns and habits of potential AI vectors. The final phase, prevent, involves the activation of deterrents. This study proposes the use of pyrotechnic sounds, human effigies, and unmanned aerial vehicles (UAVs) as potential deterrents. Pyrotechnic sounds create a sudden, loud noise that startles and scatters birds. Effigies serve as visual deterrents, exploiting birds' innate caution around perceived human presence. UAVs add a dynamic and mobile element, capable of moving and responding to bird activity in real time [56–58]. These deterrents, especially when automated, offer several advantages over traditional human-operated methods. They are not constrained by adverse weather conditions and can function in a range of temperatures, ensuring continuous protection of the farm. Moreover, they can react instantly to the presence of unwanted wild birds, providing an immediate response to potential threats.

An automated detection and deterrence system serves as a non-lethal, persistent, and adaptable solution to the problem of keeping wild birds at bay. It recognizes the impracticality of completely sealing off farms from the natural world and instead focuses on creating an environment that is unwelcoming to species that pose a risk of disease transmission. The aim is to establish a perimeter of defense that discourages wild birds from entering and lingering in areas where they could meet domestic poultry. By doing so, the system not only protects the health of the farm's birds but also preserves the ecological integrity of the local wildlife. Moreover, the continuous operation of such a system provides an ongoing dataset of bird activity, further refining the model's accuracy and the effectiveness of deterrent strategies. This allows for a proactive, rather than reactive, approach to farm management and disease prevention. Implementing an automated system also frees up resources, allowing farm personnel to focus on other critical tasks, secure in the knowledge that the AI risk is being managed efficiently. This technology-driven approach exemplifies how innovation in agricultural practices can lead to more effective, humane, and environmentally sensitive farming operations. In conclusion, combining YOLOX's detection capabilities with automated deterrents represents a modern, intelligent approach to wildlife management on farms. It is a testament to the potential of technology to enhance traditional agricultural practices, ensuring the safety and productivity of the industry while maintaining harmony with the surrounding ecosystem.

4. Conclusions

This study compares two different deep learning algorithms, i.e., YOLOX and YOLOv5, to develop wild bird detection models. The overall mAP of the models ranges from of 86.4% to 95.4% (when the IoU threshold is 0.5), indicating that the deep learning method is a useful approach to monitor wild birds. In addition, YOLOX is superior to YOLOv5 as it improves the mAP by 9% using the same training dataset, and it could be employed as the final model for wild bird detection. Some common wild birds (i.e., the gadwall and brown thrasher) can be recognized by our model, and their behavior patterns were analyzed for their application in a deterrent system to prevent AI transmission. This method plays a crucial role in developing a computer vision-based system to accurately locate potential wild birds that are reservoirs of AI viruses, which can reduce the risk of AI transmission from wild birds to poultry houses. However, this study has some limitations. Detection accuracy may be affected by weather conditions and occlusions. The model also relies on high-quality annotated data and larger, more diverse datasets are needed for better generalization. Moreover, further developments could integrate this method with unmanned aerial vehicles to automatically improve biosecurity on poultry farms, especially in open-field areas.

Author Contributions: Conceptualization, L.C.; Methodology, X.Y. and L.C.; Validation, X.Y.; Formal analysis, X.Y. and Z.W.; Investigation, X.Y., R.B.B., S.S., B.P. and L.C.; Resources, L.C.; Writing—original draft, X.Y., R.B.B., S.S., Z.W., T.L., B.P. and L.C.; Supervision, T.L.; Project administration, L.C. All authors have read and agreed to the published version of the manuscript.

Funding: This study was funded by the University of Georgia (UGA) Provost Office Rural Engagement Grant; USDA-NIFA AFRI (2023-68008-39853); Georgia Research Alliance (Venture Fund); Oracle America (Oracle for Research Grant, CPQ-2060433); and UGA IIPA Equipment grant.

Data Availability Statement: The datasets generated, used, and/or analyzed during the current study will be available from the corresponding author on reasonable request.

Conflicts of Interest: The authors declare no conflict of interest.

References

1. Zhao, Y.; Richardson, B.; Takle, E.; Chai, L.; Schmitt, D.; Xin, H. Airborne Transmission May Have Played a Role in the Spread of 2015 Highly Pathogenic Avian Influenza Outbreaks in the United States. *Sci. Rep.* **2019**, *9*, 11755. [CrossRef] [PubMed]
2. CDC. Avian Influenza Current Situation Summary. Available online: https://www.cdc.gov/bird-flu/situation-summary/index.html (accessed on 11 March 2023).
3. Chai, L.; Zhao, Y.; Xin, H.; Richardson, B. Heat Treatment for Disinfecting Egg Transport Tools. *Appl. Eng. Agric.* **2022**, *38*, 343–350. [CrossRef]
4. Krauss, S.; Webster, R.G. Avian Influenza Virus Surveillance and Wild Birds: Past and Present. *Avian Dis.* **2010**, *54*, 394–398. [CrossRef] [PubMed]
5. Vaidya, N.K.; Wang, F.-B.; Zou, X. Avian Influenza Dynamics in Wild Birds with Bird Mobility and Spatial Heterogeneous Environment. *DCDS-B* **2012**, *17*, 2829–2848. [CrossRef]
6. Comin, A.; Klinkenberg, D.; Marangon, S.; Toffan, A.; Stegeman, A. Transmission Dynamics of Low Pathogenicity Avian Influenza Infections in Turkey Flocks. *PLoS ONE* **2011**, *6*, e26935. [CrossRef]
7. Lupiani, B.; Reddy, S.M. The History of Avian Influenza. *Comp. Immunol. Microbiol. Infect. Dis.* **2009**, *32*, 311–323. [CrossRef]
8. Martinez, L.; Cheng, W.; Wang, X.; Ling, F.; Mu, L.; Li, C.; Huo, X.; Ebell, M.H.; Huang, H.; Zhu, L.; et al. A Risk Classification Model to Predict Mortality Among Laboratory-Confirmed Avian Influenza A H7N9 Patients: A Population-Based Observational Cohort Study. *J. Infect. Dis.* **2019**, *220*, 1780–1789. [CrossRef]
9. Bouma, A.; Claassen, I.; Natih, K.; Klinkenberg, D.; Donnelly, C.A.; Koch, G.; Boven, M. van Estimation of Transmission Parameters of H5N1 Avian Influenza Virus in Chickens. *PLoS Pathog.* **2009**, *5*, e1000281. [CrossRef]
10. Poulson, R.L.; Brown, J.D. Wild Bird Surveillance for Avian Influenza Virus. In *Animal Influenza Virus: Methods and Protocols*; Methods in Molecular Biology; Spackman, E., Ed.; Springer: New York, NY, USA, 2020; pp. 93–112. ISBN 978-1-07-160346-8.
11. Kandeil, A.; Patton, C.; Jones, J.C.; Jeevan, T.; Harrington, W.N.; Trifkovic, S.; Seiler, J.P.; Fabrizio, T.; Woodard, K.; Turner, J.C.; et al. Rapid Evolution of A(H5N1) Influenza Viruses after Intercontinental Spread to North America. *Nat. Commun.* **2023**, *14*, 3082. [CrossRef]
12. Li, C.; Peng, Q.; Wan, X.; Sun, H.; Tang, J. C-Terminal Motifs in Promyelocytic Leukemia Protein Isoforms Critically Regulate PML Nuclear Body Formation. *J. Cell Sci.* **2017**, *130*, 3496–3506. [CrossRef]
13. Li, C.; Fu, J.; Shao, S.; Luo, Z.-Q. Legionella Pneumophila Exploits the Endo-Lysosomal Network for Phagosome Biogenesis by Co-Opting SUMOylated Rab7. *PLoS Pathog.* **2023**, *20*, e1011783. [CrossRef] [PubMed]
14. Balaji, V.S.; Mahi, A.R.; Anirudh Ganapathy, P.S.; Manju, M. Scarecrow Monitoring System: Employing Mobilenet Ssd for Enhanced Animal Supervision. *arXiv* **2024**, arXiv:2407.01435.
15. Maheswaran, S.; Ramya, M.; Priyadharshini, P.; Sivaranjani, P. A Real Time Image Processing Based System to Scaring the Birds from the Agricultural Field. *Indian J. Sci. Technol.* **2016**, *9*, 98999. [CrossRef]
16. Ge, Y.; Yao, Q.; Wang, X.; Chai, H.; Deng, G.; Chen, H.; Hua, Y. Detection of Reassortant Avian Influenza A (H11N9) Virus in Wild Birds in China. *Transbound. Emerg. Dis.* **2019**, *66*, 1142–1157. [CrossRef] [PubMed]
17. Styś-Fijoł, N.; Kozdruń, W.; Czekaj, H. Detection of Avian Reoviruses in Wild Birds in Poland. *J. Vet. Res.* **2017**, *61*, 239–245. [CrossRef] [PubMed]
18. Zhang, Y.; Li, M.; Ma, X.; Wu, X.; Wang, Y. High-Precision Wheat Head Detection Model Based on One-Stage Network and GAN Model. *Front. Plant Sci.* **2022**, *13*, 787852. [CrossRef]
19. Datar, P.; Jain, K.; Dhedhi, B. Detection of Birds in the Wild Using Deep Learning Methods. In Proceedings of the 2018 4th International Conference for Convergence in Technology (I2CT), Mangalore, India, 27–28 October 2018; pp. 1–4.
20. Yang, X.; Bahadur Bist, R.; Paneru, B.; Liu, T.; Applegate, T.; Ritz, C.; Kim, W.; Regmi, P.; Chai, L. Computer Vision-Based Cybernetics Systems for Promoting Modern Poultry Farming: A Critical Review. *Comput. Electron. Agric.* **2024**, *225*, 109339. [CrossRef]
21. Subedi, S.; Bist, R.; Yang, X.; Chai, L. Tracking Pecking Behaviors and Damages of Cage-Free Laying Hens with Machine Vision Technologies. *Comput. Electron. Agric.* **2023**, *204*, 107545. [CrossRef]
22. Yang, X.; Chai, L.; Bist, R.B.; Subedi, S.; Guo, Y. Variation of Litter Quality in Cage-Free Houses during Pullet Production. In Proceedings of the 2022 ASABE Annual International Meeting, Houston, TX, USA, 17–20 July 2022; American Society of Agricultural and Biological Engineers: St. Joseph, MI, USA, 2022.

23. Bist, R.B.; Subedi, S.; Chai, L.; Yang, X. Ammonia Emissions, Impacts, and Mitigation Strategies for Poultry Production: A Critical Review. *J. Environ. Manag.* **2023**, *328*, 116919. [CrossRef]
24. Subedi, S.; Bist, R.; Yang, X.; Chai, L. Tracking Floor Eggs with Machine Vision in Cage-Free Hen Houses. *Poult. Sci.* **2023**, *102*, 102637. [CrossRef]
25. Yang, X.; Chai, L.; Bist, R.B.; Subedi, S.; Wu, Z. A Deep Learning Model for Detecting Cage-Free Hens on the Litter Floor. *Animals* **2022**, *12*, 1983. [CrossRef] [PubMed]
26. Hammami, M.; Friboulet, D.; Kechichian, R. Data Augmentation for Multi-Organ Detection in Medical Images. In Proceedings of the 2020 Tenth International Conference on Image Processing Theory, Tools and Applications (IPTA), Paris, France, 09–12 November 2020; pp. 1–6.
27. Lin, S.-Y.; Li, H.-Y. Integrated Circuit Board Object Detection and Image Augmentation Fusion Model Based on YOLO. *Front. Neurorobot.* **2021**, *15*, 762702. [CrossRef] [PubMed]
28. Zhang, D.; Zhou, F. Self-Supervised Image Denoising for Real-World Images with Context-Aware Transformer. *IEEE Access* **2023**, *11*, 14340–14349. [CrossRef]
29. Zhang, D.; Zhou, F.; Jiang, Y.; Fu, Z. MM-BSN: Self-Supervised Image Denoising for Real-World with Multi-Mask Based on Blind-Spot Network 2023. In Proceedings of the IEEE/CVF Conference on Computer Vision and Pattern Recognition, Vancouver, BC, Canada, 18–22 June 2023.
30. Ge, Z.; Liu, S.; Wang, F.; Li, Z.; Sun, J. YOLOX: Exceeding YOLO Series in 2021. *arXiv* **2021**, arXiv:2107.08430.
31. Redmon, J.; Farhadi, A. YOLO9000: Better, Faster, Stronger. In Proceedings of the IEEE Conference on Computer Vision and Pattern Recognition, Honolulu, HI, USA, 21–26 July 2017.
32. Redmon, J.; Divvala, S.; Girshick, R.; Farhadi, A. You Only Look Once: Unified, Real-Time Object Detection. In Proceedings of the IEEE Conference on Computer Vision and Pattern Recognition, Las Vegas, NV, USA, 26 June–1 July 2016.
33. Kim, K.; Lee, H.S. Probabilistic Anchor Assignment with IoU Prediction for Object Detection. In *Computer Vision–ECCV 2020: 16th European Conference, Glasgow, UK, 23–28 August 2020, Proceedings, Part XXV 16*; Springer International Publishing: Cham, Switzerland, 2020.
34. Ren, J.; Wang, Z.; Zhang, Y.; Liao, L. YOLOv5-R: Lightweight Real-Time Detection Based on Improved YOLOv5. *J. Electron. Imaging* **2022**, *31*, 033033. [CrossRef]
35. Zhu, X.; Lyu, S.; Wang, X.; Zhao, Q. TPH-YOLOv5: Improved YOLOv5 Based on Transformer Prediction Head for Object Detection on Drone-Captured Scenarios. In Proceedings of the IEEE/CVF International Conference on Computer Vision, Virtual, 11–17 October 2021.
36. Panigrahi, S.; Raju, U.S.N. InceptionDepth-wiseYOLOv2: Improved Implementation of YOLO Framework for Pedestrian Detection. *Int. J. Multimed. Inf. Retr.* **2022**, *11*, 409–430. [CrossRef]
37. Xue, Y.; Ju, Z.; Li, Y.; Zhang, W. MAF-YOLO: Multi-Modal Attention Fusion Based YOLO for Pedestrian Detection. *Infrared Phys. Technol.* **2021**, *118*, 103906. [CrossRef]
38. Li, Q.; Zhang, C. Continual Learning on Deployment Pipelines for Machine Learning Systems. *arXiv* **2022**, arXiv:2212.02659.
39. Huang, Y.; Yang, X.; Guo, J.; Cheng, J.; Qu, H.; Ma, J.; Li, L. A High-Precision Method for 100-Day-Old Classification of Chickens in Edge Computing Scenarios Based on Federated Computing. *Animals* **2022**, *12*, 3450. [CrossRef]
40. Ren, X.; Zhang, W.; Wu, M.; Li, C.; Wang, X. Meta-YOLO: Meta-Learning for Few-Shot Traffic Sign Detection via Decoupling Dependencies. *Appl. Sci.* **2022**, *12*, 5543. [CrossRef]
41. Gauthier-Clerc, M.; Tamisier, A.; Cezilly, F. Sleep-Vigilance Trade-off in Gadwall during the Winter Period. *Condor* **2000**, *102*, 307–313. [CrossRef]
42. Rivers, J.W.; Sandercock, B.K. Predation by gray catbird on brown thrasher eggs. *SWNA* **2004**, *49*, 101–103. [CrossRef]
43. Li, G.; Hui, X.; Chen, Z.; Chesser, G.D.; Zhao, Y. Development and Evaluation of a Method to Detect Broilers Continuously Walking around Feeder as an Indication of Restricted Feeding Behaviors. *Comput. Electron. Agric.* **2021**, *181*, 105982. [CrossRef]
44. Li, G.; Huang, Y.; Chen, Z.; Chesser, G.D.; Purswell, J.L.; Linhoss, J.; Zhao, Y. Practices and Applications of Convolutional Neural Network-Based Computer Vision Systems in Animal Farming: A Review. *Sensors* **2021**, *21*, 1492. [CrossRef]
45. Kim, M.; Jeong, J.; Kim, S. ECAP-YOLO: Efficient Channel Attention Pyramid YOLO for Small Object Detection in Aerial Image. *Remote Sens.* **2021**, *13*, 4851. [CrossRef]
46. Li, B.; Zhang, J.; Zhang, C.; Wang, L.; Xu, J.; Liu, L. Rare Bird Recognition Method in Beijing Based on TC-YOLO Model. *Biodivers. Sci.* **2024**, *32*, 24056. [CrossRef]
47. Huang, X.; Huang, Q.; Zhang, N. Dual Fusion Paired Environmental Background and Face Region for Face Anti-Spoofing. In Proceedings of the 2021 5th Asian Conference on Artificial Intelligence Technology (ACAIT), Haikou, China, 29–31 October 2021; pp. 142–149.
48. Zhou, F.; Fu, Z.; Zhang, D. High Dynamic Range Imaging with Context-Aware Transformer. In Proceedings of the 2023 International Joint Conference on Neural Networks (IJCNN), Gold Coast, Australia, 18–23 June 2023; pp. 1–8.
49. Ju, Y.; Shi, B.; Jian, M.; Qi, L.; Dong, J.; Lam, K.-M. NormAttention-PSN: A High-Frequency Region Enhanced Photometric Stereo Network with Normalized Attention. *Int. J. Comput. Vis.* **2022**, *130*, 3014–3034. [CrossRef]
50. Wang, H.; Zhang, F.; Zhao, M.; Li, W.; Xie, X.; Guo, M. Multi-Task Feature Learning for Knowledge Graph Enhanced Recommendation. In Proceedings of the The World Wide Web Conference, San Francisco, CA, USA, 13–17 May 2019; Association for Computing Machinery: New York, NY, USA, 2019; pp. 2000–2010.

51. Dou, J.X.; Jia, M.; Zaslavsky, N.; Ebeid, M.; Bao, R.; Zhang, S.; Ni, K.; Liang, P.P.; Mao, H.; Mao, Z.H. Learning more effective cell representations efficiently. In Proceedings of the NeurIPS 2022 Workshop on Learning Meaningful Representations of Life, Virtual, 9 December 2022.
52. Dou, J.X.; Mao, H.; Bao, R.; Liang, P.P.; Tan, X.; Zhang, S.; Jia, M.; Zhou, P.; Mao, Z.H. The Measurement of Knowledge in Knowledge Graphs. In *Proceedings of the AAAI 2023 Workshop on Representation Learning for Responsible Human-Centric AI (R2HCAI)*; Association for the Advancement of Artificial Intelligence (AAAI): Washington, DC, USA, 2023.
53. Munster, V.J.; Baas, C.; Lexmond, P.; Waldenström, J.; Wallensten, A.; Fransson, T.; Rimmelzwaan, G.F.; Beyer, W.E.P.; Schutten, M.; Olsen, B.; et al. Spatial, Temporal, and Species Variation in Prevalence of Influenza A Viruses in Wild Migratory Birds. *PLoS Pathog.* **2007**, *3*, e61. [CrossRef]
54. Filaire, F.; Bertran, K.; Gaide, N.; Valle, R.; Secula, A.; Perlas, A.; Foret-Lucas, C.; Nofrarías, M.; Cantero, G.; Croville, G.; et al. Viral Shedding and Environmental Dispersion of Two Clade 2.3.4.4b H5 High Pathogenicity Avian Influenza Viruses in Experimentally Infected Mule Ducks: Implications for Environmental Sampling. *Vet. Res.* **2024**, *55*, 100. [CrossRef]
55. Bahl, J.; Pham, T.T.; Hill, N.J.; Hussein, I.T.M.; Ma, E.J.; Easterday, B.C.; Halpin, R.A.; Stockwell, T.B.; Wentworth, D.E.; Kayali, G.; et al. Ecosystem Interactions Underlie the Spread of Avian Influenza A Viruses with Pandemic Potential. *PLoS Pathog.* **2016**, *12*, e1005620. [CrossRef]
56. Levey, D.J.; Tewksbury, J.J.; Cipollini, M.L.; Carlo, T.A. A Field Test of the Directed Deterrence Hypothesis in Two Species of Wild Chili. *Oecologia* **2006**, *150*, 61–68. [CrossRef]
57. Cook, A.; Rushton, S.; Allan, J.; Baxter, A. An Evaluation of Techniques to Control Problem Bird Species on Landfill Sites. *Environ. Manag.* **2008**, *41*, 834–843. [CrossRef]
58. Wen, F.; Qin, M.; Gratz, P.; Reddy, N. OpenMem: Hardware/Software Cooperative Management for Mobile Memory System. In Proceedings of the 2021 58th ACM/IEEE Design Automation Conference (DAC), San Francisco, CA, USA, 5–9 December 2021; pp. 109–114.

AgriEngineering

MDPI

Article

ACMSPT: Automated Counting and Monitoring System for Poultry Tracking

Edmanuel Cruz [1,2,†], Miguel Hidalgo-Rodriguez [1,†], Adiz Mariel Acosta-Reyes [1], José Carlos Rangel [2,3,*], Keyla Boniche [4] and Franchesca Gonzalez-Olivardia [5]

1 Centro Regional de Veraguas, Universidad Tecnológica de Panamá, Atalaya 0901, Panama; edmanuel.cruz@utp.ac.pa (E.C.); miguel.hidalgo@utp.ac.pa (M.H.-R.); adiz.acosta@utp.ac.pa (A.M.A.-R.)
2 Sistema Nacional de Investigación (SNI), SENACYT, Panama City 0816-02852, Panama
3 Facultad de Ingeniería de Sistemas Computacionales, Universidad Tecnológica de Panamá, Panama City 0819-07289, Panama
4 Facultad de Ingeniería Mecánica, Universidad Tecnológica de Panamá, Panama City 0819-07289, Panama; keyla.boniche@utp.ac.pa
5 AirES Research Group, CINEMI, Universidad Tecnológica de Panamá, Panama City 0819-07289, Panama; franchesca.gonzalez@utp.ac.pa
* Correspondence: jose.rangel@utp.ac.pa; Tel.: +507-560-3928
† These authors contributed equally to this work.

Abstract: The poultry industry faces significant challenges in efficiently monitoring large populations, especially under resource constraints and limited connectivity. This paper introduces the Automated Counting and Monitoring System for Poultry Tracking (ACMSPT), an innovative solution that integrates edge computing, Artificial Intelligence (AI), and the Internet of Things (IoT). The study begins by collecting a custom dataset of 1300 high-resolution images from real broiler farm environments, encompassing diverse lighting conditions, occlusions, and growth stages. Each image was manually annotated and used to train the YOLOv10 object detection model with carefully selected hyperparameters. The trained model was then deployed on an Orange Pi 5B single-board computer equipped with a Neural Processing Unit (NPU), enabling on-site inference and real-time poultry tracking. System performance was evaluated in both small- and commercial-scale sheds, achieving a precision of 93.1% and recall of 93.0%, with an average inference time under 200 milliseconds. The results demonstrate that ACMSPT can autonomously detect anomalies in poultry movement, facilitating timely interventions while reducing manual labor. Moreover, its cost-effective, low-connectivity design supports broader adoption in remote or resource-limited environments. Future work will focus on improving adaptability to extreme conditions and extending this approach to other livestock management contexts.

Keywords: livestock management; poultry monitoring; edge computing; artificial intelligence (AI)

1. Introduction

As the world changes constantly, so does the way we produce our food. Recent discoveries of various technologies have driven advancements in agro-industrial processes. Among the most revolutionary technologies is edge computing, which has significantly transformed diverse industries by bringing data processing closer to the point of generation, reducing latency, and improving operational efficiency [1].

Edge computing is a data processing model that brings computation and storage closer to where the data are generated rather than relying solely on centralized data centers. This

https://doi.org/10.3390/agriengineering7030086

technology represents a significant shift in how data is managed and processed, offering faster, more efficient, and more secure solutions for various industries and applications [1].

Moreover, artificial intelligence (AI) has revolutionized the way we optimize and automate processes in the agri-food industry. Beyond being a technical tool, AI has been presented as a transformative force reshaping both the economy and social processes through the analysis of large volumes of data [2].

AI allows for complex analysis, such as in predicting weather patterns, identifying pests and diseases in crops, and enhancing agricultural yields, by optimizing resources such as water and nutrients. In 2017, Katz highlighted that the resurgence of AI is deeply tied to the use of big data and its ability to interpret metrics, redefining current technological expectations and applications. Furthermore, through the use of machine learning algorithms, AI can identify problems in real time, allowing farmers to make informed decisions and address issues before they become serious threats to production [2].

The combination of AI with sensor technologies in the Internet of Things (IoT) has revolutionized the continuous monitoring of critical variables such as soil moisture and temperature. A notable example is the system proposed by Wu et al. [3], also presented in [4–6], which employs wireless sensors, mobile applications, and advanced deep learning algorithms to measure and predict soil conditions in real time.

Based on systems such as wireless sensor networks and edge computing platforms, this approach delivers fast and accurate results directly in the field, eliminating the need to transmit large volumes of data to remote servers. Additionally, implementing strategies like deep reinforcement learning enhances forecast accuracy, optimizes efficiency, and reduces energy consumption in data processing [3].

Edge computing, AI, and IoT are redefining modern agriculture by creating an intelligent and connected ecosystem. These technologies not only enhance farmers' productivity and profitability but also promote sustainable and environmentally friendly production by minimizing resource waste and reducing environmental impact. Thus, the agriculture industry of the future is shaping up to be a dynamic, innovative, and resilient sector capable of addressing global challenges related to food security and sustainability through the strategic use of cutting-edge technologies.

Broiler farms face various challenges, with the prevention and control of diseases being among the most critical. The rapid spread of illnesses and pathogens in densely populated barns or sheds can lead to high mortality rates. Among these, ESBL-producing *E. coli*, an opportunistic pathogen, can negatively impact the broiler growth cycle and presents a serious threat to poultry production and public health through zoonotic transmission. Consequently, developing early-stage control strategies is essential [7]. The inherent susceptibility of chickens to pathogens during early growth renders appropriate interventions at this developmental phase beneficial for microbiome development [8]. Therefore, there is a pressing need to develop tools for more dynamic health and administrative management, enabling decision making during flock development based on highly reliable data. AI can provide valuable insights through the monitoring of poultry populations, allowing farm managers or those responsible for health management to take timely action.

Poultry facilities and similar hostile environments make it hard to count birds. Therefore, robust and adaptive algorithms are needed to accurately detect and track them over time, ensuring reliable data analysis [9]. Typically, poultry farmers rely on daily in-person inspections to assess broiler health, using their experience to identify sick birds, which are then manually removed to prevent disease spread [10].

As part of health and administrative management, poultry farmers estimate the number of broilers in their population at two stages in the rearing process: at the beginning, when they enter the broiler house, and at the end, when the chickens are sent to the slaughterhouse.

In response to these challenges, this study aims to develop and validate an AI-driven automated poultry flock management system utilizing the YOLOv10 model. The scientific objective of this research is to advance computer vision techniques in poultry monitoring by addressing critical limitations in real-world farm environments. By integrating YOLOv10 with edge computing, our approach enables real-time, high-accuracy detection in resource-constrained settings, distinguishing it from conventional cloud-based monitoring systems that suffer from latency and connectivity issues.

From a practical standpoint, this research introduces a scalable and cost-effective solution that facilitates automated poultry monitoring and data-driven decision making. The proposed system enhances flock management by reducing manual labor, improving tracking accuracy, and optimizing resource allocation. By mitigating economic losses and promoting sustainable agricultural practices, this study offers a practical and impactful solution to modern poultry farming challenges.

The remainder of this paper is organized as follows: Section 2 provides a comprehensive review of related work, Section 3 details the methodology, Section 4 presents the findings, Section 5 offers an in-depth discussion, and Section 6 concludes with key insights and directions for future research.

2. Background

The poultry industry is undergoing rapid transformation, driven by increasing global demand for poultry products and technological advancements that enhance production efficiency and sustainability. As global populations grow and consumer preferences shift, the poultry sector plays a central role in addressing the rising need for affordable, high-quality protein sources. However, numerous challenges persist, including with respect to mitigating environmental impacts, improving animal welfare, and ensuring food security [11,12].

To address these challenges, modern technologies are being integrated into poultry farming. Precision livestock farming (PLF) systems, utilizing sensors, cameras, and AI technologies, have emerged as key solutions for improving farm management, monitoring animal welfare, and increasing production efficiency. These systems facilitate real-time decision making, allowing farmers to undertake informed interventions before issues escalate [13,14]. Sustainability concerns have also driven efforts to reduce the environmental footprint of farming practices by minimizing water use, reducing emissions, and transitioning to more sustainable feed sources. Meanwhile, poultry welfare is being prioritized through housing systems such as by adopting free-range and organic practices, which positively impact bird health and productivity [11,15]. Research such as that presented in [16] focuses on the relationship between stocking density and gut microbiome development, with findings supporting the hypothesis that an increase in space positively correlates with improvements in weight and gut health. Despite their promise, the implementation of these technologies faces barriers, including technical constraints, high initial costs, and the need for skilled management [12,14].

The increasing adoption of the agricultural Internet of Things (IoT) has necessitated efficient data processing solutions. Traditional cloud computing systems, though effective, often suffer from latency and network congestion due to the vast amount of data generated in large agricultural areas. Edge computing addresses this challenge by processing data closer to the source, reducing latency and alleviating the load on centralized cloud servers [17,18]. In agriculture, edge computing enhances efficiency and enables immediate decision making, with applications in pest identification, product safety traceability, and agricultural machinery management [17]. By leveraging AI algorithms at the edge, real-time analysis of sensor data becomes possible, allowing for faster responses to changing farm conditions.

Artificial intelligence (AI) plays a crucial role in modern farming techniques, particularly in crop and livestock management. AI technologies, including machine learning (ML), deep learning (DL), and neural networks, contribute to the prediction of crop yields, the monitoring of plant health, and the detection of pests or diseases [19,20]. AI-powered drones and sensors provide detailed insights into soil conditions, water usage, and crop health, facilitating targeted interventions [21]. These capabilities optimize resource use, reduce waste, and enhance yield quality. As shown in [22,23], the use of computer vision-enabled robotic arms for automated decision making in vegetable and crop handling is becoming more common, applying a YOLO implementation. By analyzing data from IoT devices, AI can recommend optimal planting times, irrigation schedules, and fertilization practices, thereby promoting sustainability [24]. Furthermore, AI can predict the optimal harvesting time, maximizing productivity and minimizing losses [21].

The integration of edge computing and AI offers significant advantages in precision agriculture. By deploying AI algorithms locally on farm devices, real-time data analysis and decision making can occur without reliance on constant cloud connectivity. This integration enhances resource efficiency, improves crop management, and strengthens pest control [18,24]. For example, AI can analyze data from soil moisture sensors, weather stations, and crop health monitors at the edge, providing immediate feedback on soil conditions, irrigation needs, and pest threats. These advancements help optimize operations and reduce environmental footprints by minimizing water usage, pesticide application, and fertilizer runoff [17,24]. The combination of these technologies is transforming agriculture, providing farmers with essential tools to improve efficiency, ensure better yields, and promote sustainability.

In 2020, Ren et al. [20] analyzed agricultural robotics applications for poultry production and classified them into three functional groups according to their functions: monitors, harvesters, and combined monitor–harvesters. Ren et al. [20] reported that neural networks have been employed for image-based detection across a wide range of farm animals, including pigs, goats, and cows. The authors posited that the identified trajectory of research and development will be maintained, with poultry management efficiency representing an increasingly prominent area of scholarly inquiry.

The use of AI and computer vision in poultry farming has revolutionized monitoring and management practices. AI-based object detection models such as YOLO (You Only Look Once) [25] and Faster R-CNN [26] automate tasks like poultry counting, behavior monitoring, and health assessment. These models demonstrate high accuracy in detecting and tracking chickens, enhancing farm efficiency and resource management. AI-driven monitoring systems using mainly YOLOv5 and YOLOv8 have been particularly effective in poultry counting and health assessments.

A dead chicken detection model utilizing YOLOv4 was trained, achieving a precision of 95.24%, and a mobile robotic system was designed and constructed for the purpose of removing detected carcasses from sheds [10].

Likewise, YOLOv5 has successfully classified and counted different chicken breeds, achieving precision and recall rates of 89.9% and 87.2%, respectively, providing a practical alternative to manual counting [27]. YOLOv5 has seen widespread use in various research areas. In a study published in [28], a modified YOLOv5 model was used for chicken detection in farm settings, yielding an mAP of 90.0%. The authors also articulated the shortcomings of current detection methodologies.

Employing YOLOv5 in conjunction with a Kalman filter, researchers have explored chicken tracking to measure activity levels. This approach provides continuous monitoring of chickens' perching, walking, and environmental interactions [29].

An alternative approach to broiler weight estimation involves the integration of a YOLOv5 model and a convolutional neural network employing a regression output [30]. The system was implemented using a Raspberry Pi board and achieved a 0.87 R^2 value.

The detection of diseases utilizing the YOLOv5 architecture was detailed in [31], wherein a ResNet50 classification model was integrated to identify visual symptoms of illness in chickens, leading to a 35% improvement in their overall health.

For instance, YOLOv8 models have achieved an F1 score of 0.86, indicating high accuracy in detecting pathological features in chickens and monitoring crucial environmental parameters like temperature and humidity [4].

A YOLOv8-based approach incorporating data augmentation was used to predict pathological characteristics in chickens, yielding a test precision of 89.9% [4]. To keep owners informed, this approach employs IoT devices to send notifications. A model based on the YOLOv8 architecture was developed by Chemme and Alitappeh [32] for chicken detection within a cluttered environment, attaining an accuracy of 93%, thereby surpassing the results yielded in [29].

The reported in [9], aimed to refine automated chicken identification in videos, tackling the dynamic and inconsistent conditions of poultry farms. The YOLOv8n model was selected for its portability in chicken detection. The methodology for chicken tracking, subjected to testing with concurrent combinations of chickens, exhibited an accuracy rate of 98.3% in scenarios characterized by multiple missing and coalescing chickens.

Elmessery et al. [33] presented a new method for chicken detection, combining computer vision and YOLO-based algorithms to be applied to both thermal and visual images with diverse, complex backgrounds. The detection was focused on pathological phenomena of broilers. The authors conducted a comparative analysis of YOLO architectures, specifically v8, v7, and v5, and determined that YOLOv8 exhibited the highest performance metrics, achieving mAP50 and mAP50-95 values of 0.829 and 0.679, respectively, for visual images, demonstrating its superiority over the preceding versions.

Recognizing poultry locomotion as a key indicator of health, welfare, and productivity, Yang et al. [34] introduced an enhanced Track Anything Model (TAM) for chicken tracking. This model, achieving an RMSE of 0.02 m/s in speed detection, provides a reliable, non-intrusive method for locomotion analysis and outperforms YOLOv5 and YOLOv8 in chicken tracking. Beyond simple counting, AI-powered computer vision systems enhance monitoring by detecting behavioral anomalies and identifying sick birds. These systems integrate with IoT devices to measure environmental parameters, triggering alerts when conditions deviate from optimal ranges. This approach provides real-time, non-intrusive solutions for maintaining a stress-free and healthy environment, leading to improved productivity and welfare standards [35]. For example, smart poultry management systems based on IoT technology utilize sensors to monitor temperature, humidity, and ammonia levels, automating farm management tasks such as feeding and watering. Remote monitoring through mobile apps and websites further improves operational efficiency and ensures optimal poultry conditions [36].

Wireless sensor networks (WSNs) also play a vital role in modern poultry farm management. These networks enable continuous monitoring of environmental conditions, with real-time data transmission to centralized systems that analyze information for informed decision making. WSNs are widely used for the monitoring of temperature, humidity, and ammonia levels, ensuring optimal poultry health [37]. When integrated with AI models such as Faster R-CNN, these systems can detect dead chickens and count live ones, offering a more automated and efficient approach to farm management [38]. The combination of AI-driven object detection, IoT sensors, and WSNs represents a powerful solution for modern poultry farming, enhancing farm management, improving poultry welfare, and optimizing resource use, thereby contributing to the future of sustainable poultry farming. A selection of projects employing the YOLO architecture for model training is illustrated in Table 1. These projects concentrate on chicken tracking and detection within agricultural environments. The presented results demonstrate that YOLOv8 tends to display superior performance when compared to prior versions.

Table 1. Models trained using the YOLO architecture for problems related to chicken detection on farms. NA stands for not available data about this metric in the study.

Models	Precision	Recall	F1-Score	mAP@0.5	mAP@0.5:95	YOLO Version	Approach
Liu et al. [10]	95.24	100	NA	100	NA	v4	Dead chicken detection
Sun et al. [27]	89.9	87.2	NA	NA	NA	v5	Detection
Zhang et al. [28]	NA	NA	NA	90	62.7	v5	Detection
Chemme and Alitappeh [32]	87	85	NA	90	82	v8	Detection
Paramathma et al. [4]	93.8	77.4	86	82.1	NA	v8	Disease detection
Mehdizadeh et al. [9]	98.3	99.2	99.2	NA	NA	v8	Detection and tracking
Elmessery et al. [33]	86.1	69.1	76.7	82.9	67.9	v8	Detection
Cruz et al. [35]	93.1	93	91	93.1	62.5	v8	Detection

The review provided in [39] examined the diverse applications of the YOLO architecture throughout its development. The study pointed out the different versions of YOLO used in areas such as medicine, autonomous driving, industry, and agriculture. It stated that, specifically in agriculture, YOLOv8 has achieved the most favorable results. Likewise [40] concentrated on analyzing the use of YOLO within agriculture, particularly in crop cultivation. The authors detailed the architectural evolution and the enhancements achieved with each subsequent version. Considering this information and the outcomes of earlier studies [35], the decision was made to utilize the YOLOv10 architecture, representing the most current version during the development phase. An evaluation of version enhancements, which have consistently yielded better results, justifies this determination. Furthermore, Section 3.2.4 provides evidence that YOLOv10 is compatible with the project's chosen hardware.

3. Materials and Methods

3.1. Research Methodology

3.1.1. Description of the Approach

This study employs an applied research approach to develop an automated poultry monitoring system using edge computing and artificial intelligence (AI). The system is designed to address the challenges associated with the monitoring of large-scale poultry populations in dynamic farm environments, which include variations in lighting conditions, occlusions, and infrastructure constraints. By leveraging edge computing, data processing is performed locally, reducing latency and dependence on cloud-based solutions—an essential feature for farms with limited network connectivity [41].

To achieve real-time monitoring, an object detection model based on the YOLO (You Only Look Once) architecture was deployed on a low-power single-board computer (SBC), an Orange Pi 5B. This device, equipped with a neural processing unit (NPU), enables efficient on-device AI inference. The YOLO-based model has been widely recognized for its superior performance in object identification and counting tasks within agricultural monitoring applications [25,42].

The integration of edge computing and AI aligns with current advancements in precision agriculture, where real-time decision making is critical for livestock management. Studies have demonstrated that Edge Computing enhances autonomy and reduces operational latency, making it particularly effective for the monitoring of poultry farms [43,44]. Furthermore, automated detection of behavioral patterns, such as activity levels and mobility changes, enables early interventions to mitigate disease outbreaks or stress-induced anomalies [44].

Given the complexity of farm environments, robust object detection models such as YOLO have proven effective for real-time image and video analysis [42]. The YOLO series is well-suited for high-speed applications, processing entire video frames in a single pass. This makes it an ideal candidate for the tracking and counting of poultry populations with high accuracy. Additionally, deploying YOLO models on energy-efficient single-board computers, such as an Orange Pi with an NPU, optimizes performance while maintaining low power consumption—an essential factor for sustainable and resource-constrained agricultural settings.

3.1.2. Tool Selection Process

The selection of hardware and software for this project was based on efficiency, scalability, and suitability for real-time monitoring in environments with limited network access. A key requirement was ensuring that the system could operate autonomously with minimal reliance on external infrastructure [41].

Among various single-board computer (SBC) options, the Orange Pi 5B was chosen due to its Rockchip RK3588S processor, which integrates an eight-core CPU with a neural processing unit (NPU) capable of 6 TOPS (Tera Operations Per Second). This hardware enables low-latency inference of complex models such as YOLOv10, making it ideal for edge computing applications on rural poultry farms, where speed and autonomy are critical [45]. Furthermore, its architecture supports multiple sensors and cameras, allowing for continuous real-time data collection [45].

For object detection, the YOLOv10 (You Only Look Once) model was selected due to its superior balance of accuracy and computational efficiency in agricultural monitoring applications. YOLOv10 efficiently processes video frames in real time, making it well-suited for the tracking and counting of moving poultry. Alternative models, such as SSD and Faster R-CNN, were considered but not chosen due to their higher computational demands, which could reduce performance in resource-constrained environments [46].

To ensure high-quality training data, labelImg was used as the annotation tool due to its widespread adoption in the computer vision community and its ability to facilitate precise segmentation [47]. This step was crucial in maintaining consistency in dataset labeling, particularly under challenging conditions such as occlusion and variable lighting [48]. Additionally, a manual review of annotations was conducted to further enhance label accuracy.

The selection process prioritized adaptability and scalability. The combination of the Orange Pi 5B and YOLOv10 allows for autonomous, real-time monitoring with minimal connectivity requirements. This design enables poultry farmers to quickly detect changes in bird activity and population dynamics, ultimately improving efficiency and decision making in farm management [46].

3.1.3. Dataset Creation

The dataset was developed to ensure the YOLOv10 detection model could accurately and reliably operate in poultry farm environments. Real-time images were captured from a broiler chicken farm using ceiling-mounted cameras to cover the entire rearing area. Over six weeks, 1300 high-resolution images (1920 × 1080 pixels) were collected, representing different chicken growth stages, activities, lighting conditions, and occlusions.

Figure 1 presents sample images from the dataset, categorized by location: the left column corresponds to the small shed, while the right column represents the commercial shed.

The dataset encompasses various growth phases: the initial stage (Figure 1a,b) shows less clutter, the middle stage (Figure 1c,d) exhibits increased clustering around feeders, and the final stage (Figure 1e,f) captures fully grown chickens in a crowded environment.

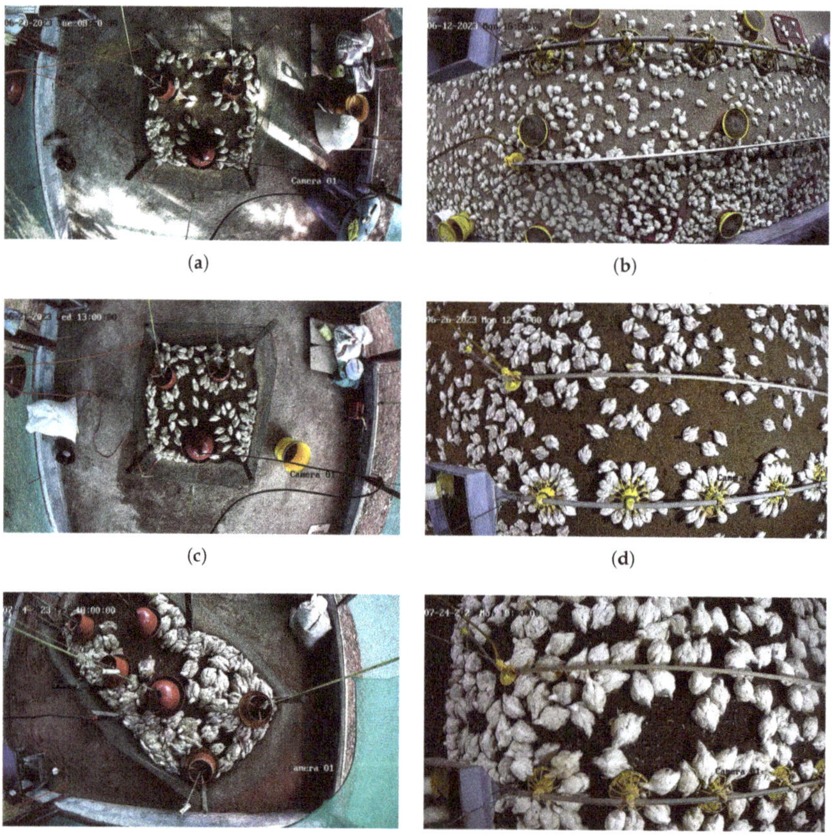

(a) (b) (c) (d) (e) (f)

Figure 1. Sample images of the dataset, depicting different chicken growth stages in the small shed (**left column**) and commercial shed (**right column**). The first row (**a**,**b**), captures the early growth stage. The middle row, (**c**,**d**) illustrates mid-stage clustering. The bottom row, (**e**,**f**) shows the fully grown stage with increased crowding and lighting variations.

To replicate real farm conditions, image collection included both natural and artificial lighting, as well as shadowed areas, capturing the daily behavioral variations of chickens. This diversity ensures the AI model's robustness in handling varying environmental factors.

All images were manually annotated using the labelImgtool [47], which is widely used in computer vision for precise dataset labeling. Each chicken was marked with a bounding box to ensure accurate detection, even under occlusion and varying viewing angles. To enhance labeling consistency, each annotation was reviewed by at least two individuals based on predefined criteria, ensuring only visible chickens were marked with precise boundaries [48].

For model training, the dataset was split into two subsets: 70% for training and 30% for testing and validation. This allocation ensures that the model learns from diverse scenarios while being evaluated on unseen data to assess its generalization ability. Additionally, data augmentation techniques such as rotation, scaling, cropping, and brightness/contrast adjustments were applied to improve model robustness to environmental variations.

The accuracy of the YOLOv10 model relies heavily on the quality of the dataset. This carefully curated dataset enhances detection and counting precision while maintaining adaptability for real-world deployment. Furthermore, the dataset is expandable for future research, enabling applications in other livestock monitoring scenarios.

3.1.4. Structure of the Poultry Sheds

The study was conducted in two different poultry shed environments: a small shed housing 100 chickens and a commercial-sized shed. In the commercial shed, data collection was limited to a designated quarter of the facility, as illustrated in Figure 2a,b. To contextualize the study area, Figure 2b presents a Google Earth image of the shed, where the red square marks the specific region used for image acquisition. The commercial shed features a robust metal-frame structure, equipped with ventilation systems and optimized lighting conditions to ensure effective monitoring of a large poultry population.

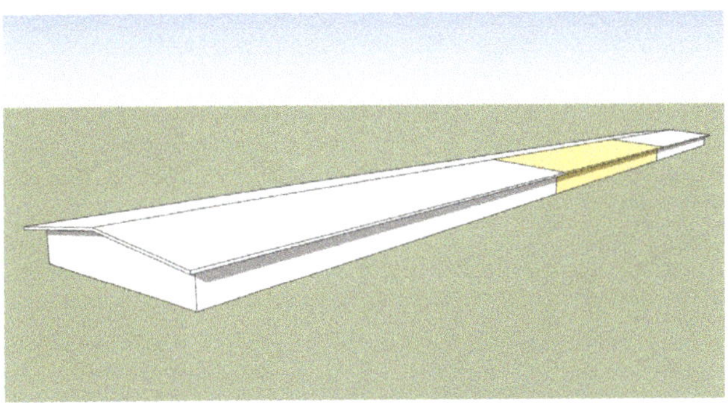

(**a**) Commercial poultry shed.

(**b**) Section of the shed used for data capture.

Figure 2. Schematic views of the commercial poultry shed used during data collection and model testing.

Conversely, the small shed (Figure 3a,b) housed a smaller flock of 100 chickens and was designed for small-scale operations. This environment featured a compact layout with a simplified ventilation and lighting system, making it suitable for detailed behavioral observations. The shed's smaller dimensions facilitated close monitoring of individual chickens but also posed certain spatial limitations.

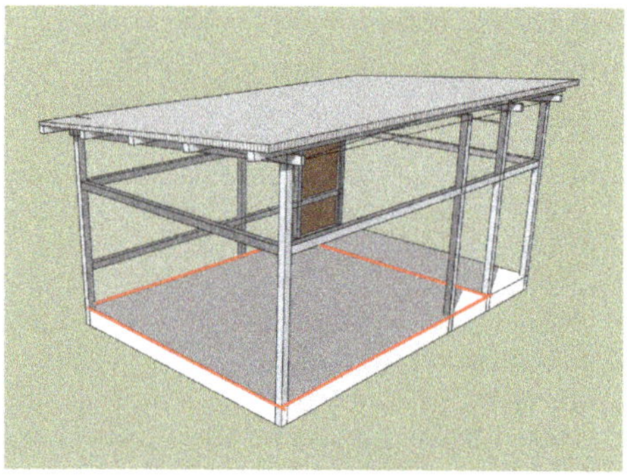

(**a**) Small poultry shed. The red area indicates the designated growing space.

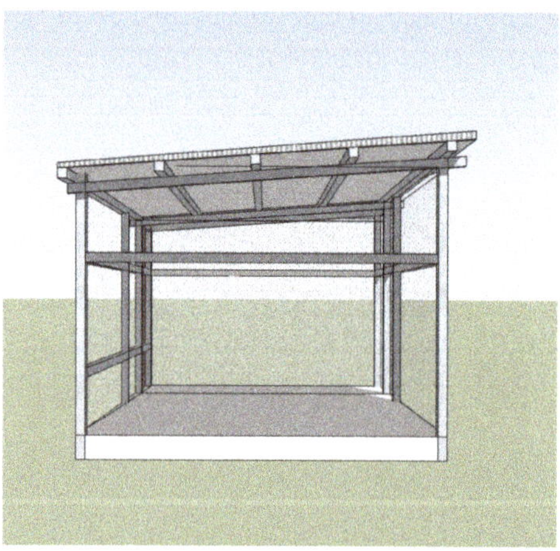

(**b**) Lateral view of the small poultry shed.

Figure 3. Schematic views of the small poultry shed used during data capture and model testing.

These two environments (Figures 2 and 3) provided diverse datasets, which were essential for the training of the YOLOv10 detection model to function effectively under different poultry farm conditions.

Figure 4 presents images from the interior of the commercial shed. In Figure 4a, red squares indicate the positions of the installed cameras used for data collection. Figure 4b provides an overview of the camera arrangement on the ceiling, while Figure 4c offers a close-up view of a single camera setup.

(a) The interior of the commercial shed, with red squares indicating camera positions.

(b) Camera arrangement on the shed ceiling.

(c) Camera configuration for image acquisition.

Figure 4. Interior of the commercial shed used for data capture.

3.2. Project Architecture

3.2.1. General Description

The system architecture follows the edge computing paradigm, ensuring all data processing occurs locally, on-site at the poultry farm. This eliminates reliance on external servers, which is a crucial feature for rural environments where Internet connectivity is often limited or unstable. The system is designed to operate autonomously and efficiently, with the Orange Pi 5B serving as the central processing unit. Equipped with a 6-TOPS (Tera operations per second) neural processing unit (NPU), this device enables low-latency, real-time inference, ensuring continuous and accurate monitoring of poultry activity and population counts.

The hardware infrastructure includes high-resolution surveillance cameras mounted on the shed ceilings, covering the entire breeding area. These cameras, capable of capturing images at a 1920 × 1080 pixel resolution, are connected to a Network Video Recorder (NVR), which centralizes and manages video streams. The Orange Pi 5B, linked to the NVR, processes these images using the YOLOv10 object detection model. The model, pre-optimized on a GPU workstation, is adapted for efficient execution on the device's NPU, enabling fast on-site analysis.

The YOLOv10 model identifies and counts birds in real time, distinguishing between active and resting chickens. This capability is critical for flock health and welfare monitoring, enabling early detection of potential stress indicators, anomalies, or disease outbreaks. By performing on-device processing, the system provides instant responses, minimizes cloud dependency, enhances data privacy, and reduces bandwidth usage.

In addition to its hardware components, the architecture includes a Web-based user interface accessible from any device within the farm's local network. This interface provides real-time visualizations of detected chickens, their activity status, and historical data analytics. The system also incorporates time-series analysis, allowing for the early detection of behavioral anomalies that may indicate potential health issues within the flock.

By integrating optimized hardware and software, this scalable and cost-effective solution enables efficient adaptation to varying farm conditions. The system maximizes productivity, reduces operational costs, and ensures autonomous monitoring through an edge computing approach tailored for real-world poultry farming challenges.

3.2.2. System Proposal

The proposed system integrates optimized hardware and software to enable real-time monitoring and counting of chicken activity. Built on an edge computing architecture, the solution deploys an artificial intelligence model on the Orange Pi 5B, ensuring autonomous on-site processing. Each system component is detailed below.

- **Hardware Configuration**
  The core hardware consists of the Orange Pi 5B, powered by the Rockchip RK3588S octa-core processor and a 6-TOPS neural processing unit (NPU). This configuration enables efficient AI inference while maintaining low power consumption. The on-device processing capability of the NPU eliminates reliance on external servers, making it ideal for environments with limited connectivity or where data privacy is a key concern.
  The system also incorporates high-resolution surveillance cameras mounted on the poultry shed ceilings to capture flock images at a 1920 × 1080 pixel resolution. These cameras are connected directly to the Orange Pi 5B via dedicated input ports, ensuring continuous real-time data acquisition across the entire breeding area.
- **Software Configuration**
  The software framework is built around YOLOv10, an advanced object detection model that balances speed and precision for real-time monitoring tasks. Initially trained on a GPU workstation, the model is subsequently optimized and deployed on the Orange Pi 5B for efficient on-site inference.
  YOLOv10 is configured in its small variant, which optimizes computational efficiency while maintaining high detection accuracy. This variant is specifically suited for resource-constrained edge devices, such as the Orange Pi, which requires rapid and continuous inferences to effectively monitor poultry activity in real time.

3.2.3. Model Training

The YOLOv10 model was trained using a dataset of 1300 images captured from the farm, each pre-labeled with bounding boxes to identify individual chickens. To optimize model performance, several key hyperparameters were employed, as summarized in Table 2.

- **Epochs**: The model was trained for 100 epochs, a value determined through preliminary experiments to ensure convergence without overfitting.

- **Batch Size**: A batch size of 16 images per iteration was chosen to balance GPU memory utilization and maintain stable training.
- **Learning Rate**: The initial learning rate was set to 0.001, with a gradual decay scheduler to refine model weights and enhance convergence.
- **Optimizer**: AdamW was selected for its efficiency in computer vision tasks and its ability to handle large datasets with precise, adaptive weight updates.
- **Data Augmentation**: To improve the model's generalization capability, various data augmentation techniques were applied, including rotation, scaling, cropping, brightness, and contrast adjustments. These augmentations simulate diverse lighting and occlusion conditions, ensuring robustness in real-world environments.

Table 2. Training hyperparameters for the YOLOv10 model.

Hyperparameter	Value
Epochs	100
Batch Size	16
Learning Rate	0.001
Optimizer	AdamW
Data Augmentation	Rotation, scaling, cropping, brightness, and contrast adjustments

Summary of the hyperparameters used for YOLOv10 model training.

3.2.4. Deployment on the Orange Pi 5B

After training, the YOLOv10 model was deployed on the Orange Pi 5B, where the neural processing unit (NPU) was configured for real-time inference. To ensure compatibility, the model was first converted to the ONNX format. This conversion addresses the operational compatibility between the original YOLOv10 and the Rockchip NPU integrated on the Orange Pi 5B board. The modifications include replacing operations that were not recognized by the compiler and transforming the single output into three separate outputs. After conversion, the ONNX model was optimized using the rknn-toolkit (https://github.com/airockchip/rknn-toolkit2/, accessed on 30 October 2024) to generate the RKNN format, which is specifically designed for the RK3588S processor's NPU architecture. The aim of this optimization is to decrease model weight, increase inference speed, and refine convolutional layers and activation functions for optimal inference efficiency. The official model documentation provides support for these procedures [49–53]. Figure 5 illustrates this conversion process.

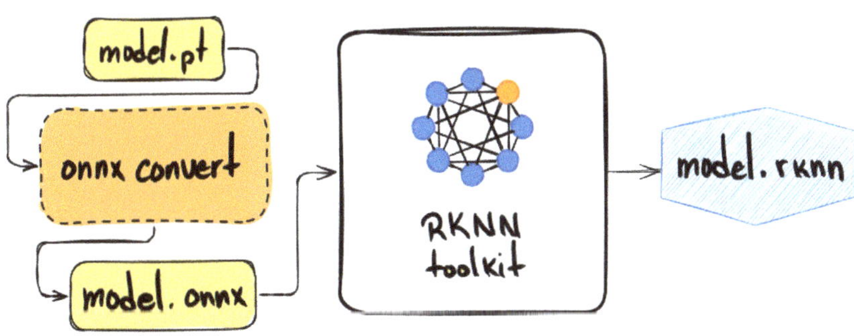

Figure 5. Diagram of the model conversion process.

Leveraging the Orange Pi 5B's efficient on-device processing, the system can handle multiple images per second and provide real-time responses to variations in flock activity.

This setup enables the detection of movement and inactivity patterns, allowing operators to receive alerts for abnormal behaviors and intervene promptly when necessary.

3.2.5. System Microservices

The system architecture is divided into four key microservices, each handling a specific function:

1. **Image Processing and Inference Microservice:** This microservice receives images from the surveillance cameras, applies preprocessing steps (e.g., color normalization and resizing), and executes YOLOv10 inference (https://docs.ultralytics.com/es/models/yolov10/, accessed on 5 October 2024) for chicken counting and motion detection. The processed data are then filtered and consolidated before being passed to the backend.
2. **Backend Microservice for Data Management:** Acting as an intermediary between the inference microservice and the database, this component receives and processes detection results, forwarding them to a MongoDB database (https://www.mongodb.com/es, accessed on 17 August 2024). Additionally, it manages business logic, including real-time alert generation and notifications for the Grafana dashboard.
3. **Grafana Microservice for Data Visualization:** Responsible for managing the Grafana instance (https://grafana.com/, accessed on 21 June 2024), this microservice extracts stored data from MongoDB and visualizes real-time metrics as interactive graphs and dashboards. Users can monitor chicken activity and receive visual alerts for unusual behaviors.
4. **Database Microservice (MongoDB):** MongoDB serves as the primary data storage system, chosen for its scalability and flexibility in handling unstructured data. It ensures the persistence of historical records, facilitates fast queries, and supports efficient trend analysis over time.

3.2.6. Docker for Container Management

To ensure portability and replicability, all microservices are encapsulated within Docker containers (https://www.docker.com/, accessed on 13 June 2024). Docker simplifies deployment across different environments by maintaining consistent dependencies and configurations. This modular approach is particularly beneficial for resource-constrained agricultural settings, allowing the entire system to run on a single Orange Pi 5B unit while ensuring scalability.

3.2.7. Data Visualization in Grafana

The Grafana interface provides real-time visual monitoring of key poultry parameters, including the following:

- Number of chickens detected;
- Motion status (active vs. inactive birds);
- Average movement speed.

The dashboard also features line and bar charts displaying behavioral trends over time, helping farmers identify potential health issues. Automated alerts notify operators of abnormal inactivity levels or sudden changes in movement patterns, enabling quick intervention when necessary.

3.2.8. Access to Historical Data and Automated Reports

Grafana, integrated with MongoDB, allows for quick access to historical data, supporting long-term behavioral analysis. Users can generate automated reports at configurable intervals (daily, weekly, or monthly) and export them in formats such as CSV or PDF.

This feature facilitates external data analysis and integration with other farm management systems, providing a comprehensive monitoring solution.

3.2.9. User Interface

The system features a user interface accessible via the local network, enabling operators to monitor real-time poultry activity, including chicken count and movement status, as well as access historical data. Designed with an intuitive dashboard, the interface incorporates charts and visual alerts to facilitate quick and informed decision making for flock management.

As illustrated in Figure 6, the system architecture consists of three main components:

- **Image Capture System**: Surveillance cameras continuously capture poultry activity.
- **Processing Unit (Orange Pi 5B)**: The YOLOv10 model performs real-time object detection and data analysis.
- **Dashboard for Visualization**: Displays real-time insights using Grafana.

The interface is implemented using Grafana, providing real-time visualization and historical analysis. Operators can access the dashboard from any device on the local network, enabling centralized monitoring of chicken population dynamics and behavioral trends on the farm.

The system infrastructure follows a microservices architecture, with all services encapsulated in Docker containers to enhance scalability, portability, and dependency management. This design ensures the system remains modular and easy to manage, allowing for flexible deployment across different farm setups.

The integration of Grafana and MongoDB delivers a comprehensive, user-friendly experience, supporting both real-time monitoring and historical analysis of poultry activity, providing valuable insights for farm management.

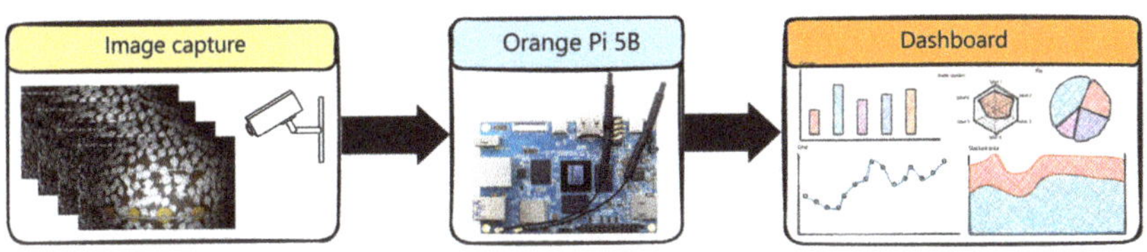

Figure 6. System architecture: Image capture, processing via Orange Pi 5B, and data visualization on the dashboard.

4. Results

4.1. Workflows

Testing Process

The system's testing process was designed to verify the functionality and accuracy of the AI model, as well as the integration and communication between the various microservices that comprise the system's architecture. This process included a series of steps for model inference and evaluation, as well as the testing of each microservice under simulated production conditions. Below are the key steps and components:

1. Preprocessing, Inference, and Postprocessing Microservice:
   This microservice performs three integrated functions:
   - **Preprocessing:** Upon receiving an image captured by the cameras, the microservice conducts initial preprocessing, including color normalization, resizing, and formatting of the image to ensure that the input data for the YOLOv10 model are consistent and optimized.

- **Inference:** The preprocessed image is sent to the YOLOv10 model, which performs real-time object detection, identifying and counting chickens in the capture area.
- **Postprocessing:** Once the inference results are obtained, the microservice applies postprocessing to consolidate the information, removing redundant detections and adjusting the bounding-box coordinates. Additionally, extra metrics are calculated, such as movement status (active or inactive chickens) and average movement speed.

This microservice ensures that the data are fully prepared before being sent to the backend for storage and visualization.

2. Backend Microservice (API):
The back end functions as an API service responsible for receiving the postprocessed data from the image processing microservice and sending them to the MongoDB database. This service acts as an intermediary, connecting the image processing microservice with the storage system, ensuring smooth and real-time data communication. The backend also manages the business logic necessary to store the data correctly, making it accessible later for visualization in Grafana.
3. Database Microservice (MongoDB):
MongoDB stores all the postprocessed data it receives from the backend. This database not only saves real-time information but also facilitates access to historical data. The data structure in MongoDB is designed to support fast and efficient queries, allowing for optimal integration with Grafana, which is used for visual data representation.
4. Visualization Microservice (Grafana):
Grafana is responsible for displaying the results stored in MongoDB, providing a complete visual interface for the user. Through interactive dashboards and charts, Grafana enables farm operators to monitor real-time chicken counts, movement status, and other relevant metrics. It also offers the ability to generate visual alerts when values exceed certain thresholds and allows for access to historical data for analysis of flock activity patterns.

4.2. *Performance Metrics*

The performance of the proposed model is evaluated using multiple metrics to demonstrate its robustness and efficiency under various conditions.

4.2.1. F1–Confidence Curve

The analysis of the F1–confidence curve demonstrated an optimal model performance with an average F1 score of 0.90 at a confidence threshold of 0.328. This result highlights the model's ability to maintain a proper balance between precision and recall in real-time inferences (Figure 7a).

4.2.2. Precision–Confidence Curve

The model's performance in terms of precision reached a maximum value of 1.00 at a confidence threshold of 0.937, indicating the system's ability to correctly identify detected objects without compromising prediction quality (Figure 7b).

4.2.3. Precision–Recall Curve

The analysis of the precision–recall curve reflects an average precision of 0.928 at a 50% threshold, validating the model's robustness in handling data with variability in capture conditions (Figure 7c).

4.2.4. Recall–Confidence Curve

The recall–confidence curve showed an average recall of 0.92, with an optimal threshold at 0. This result indicates the system's ability to minimize false negatives, detecting most of the objects of interest present in the images (Figure 7d).

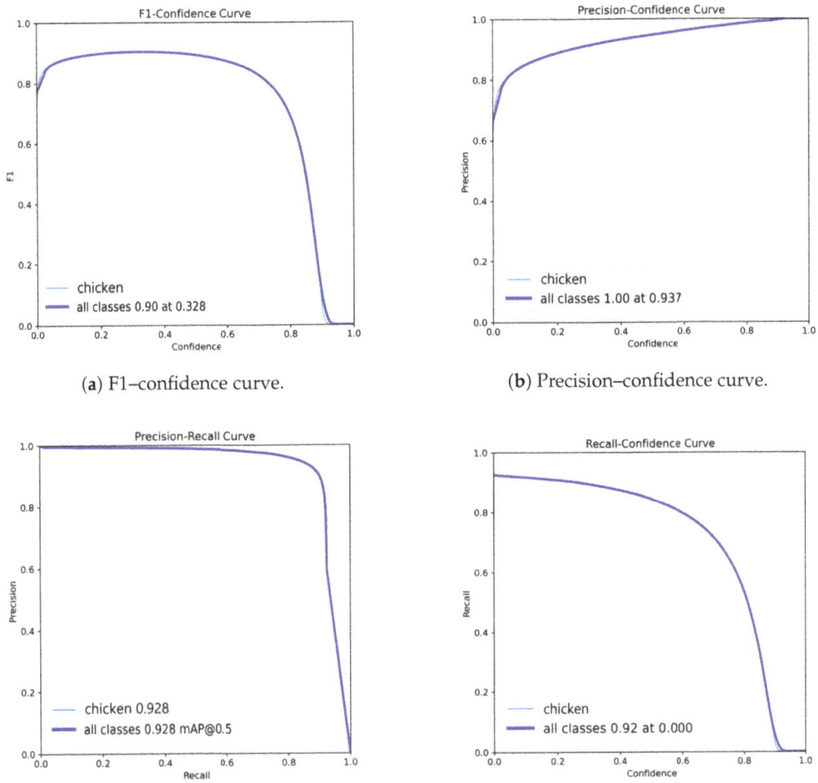

(**a**) F1–confidence curve. (**b**) Precision–confidence curve.

(**c**) Precision–recall curve. (**d**) Recall–confidence curve.

Figure 7. Visual representation of the performance metrics of the proposed model: (**a**) F1–confidence curve, showing an average F1 score of 0.90 at a confidence threshold of 0.328; (**b**) precision–confidence curve, achieving a maximum precision of 1.00 at a confidence threshold of 0.937; (**c**) precision–recall curve, indicating an average precision of 0.928 at a 50% threshold; (**d**) recall–confidence curve, with an average recall of 0.92 and an optimal threshold of 0.

4.2.5. Real-Time Detection Visualization

In addition to numerical metrics, Figure 8 demonstrates the model's real-time detection capability in a complex barn environment. Bounding boxes represent identified poultry with associated confidence scores, illustrating the model's practical implementation and its ability to handle variable conditions such as changes in lighting and occlusion.

The YOLOv10 model demonstrated a precision of 93.1% and recall of 93.0% under real-time farm conditions, with an average inference time below 200 milliseconds. Additionally, during a test run of 5 min, the system processed 2359 images. This translates to the following metrics:

- **Average Inference Time per Image**: 0.0491 s.
- **Inference Speed**: 20.35 FPS (Frames Per Second).

These results confirm the efficiency of the YOLOv10 model in real-time poultry detection and counting tasks, meeting the demands of low-latency, high-accuracy requirements in agricultural environments.

Figure 8. Example of poultry detection using the YOLOv10 model in real time. Bounding boxes and associated confidence scores highlight the model's successful identification and monitoring of individual poultry in a complex farm environment.

4.2.6. Real-World Testing

Real-world tests were conducted for one hour with a camera installed in a commercial poultry barn. During this period, the system detected 1548 chickens, corresponding to the birds that entered the camera's field of view. These results demonstrate the system's effectiveness in an operational environment, under real-world lighting and movement conditions. The YOLOv10 detection model's ability to accurately identify and count the birds remained high, showcasing the system's robustness under changing conditions.

The tests validated the model's accuracy for automatic bird counting, a key feature of the system that enables efficient and precise management of poultry populations in real time, minimizing the need for manual intervention. The system is designed to operate within a closed, isolated network, disconnected from the Internet due to the geographical constraints of the available poultry shed. This isolation provides a fundamental level of security by restricting access to only those devices that are physically connected to the network. For deployment in varied locations, a careful evaluation is crucial to implement security and data access strategies that achieve a balance among usability, efficiency, and safeguarding against unauthorized access.

Figure 9 presents the results panel and the hardware setup used during testing. Figure 9a displays the captured image with labeled chickens alongside the Orange Pi's operational status at the time of execution. Figure 9c provides an overview of the complete system, while Figure 9b offers a detailed view of the hardware components. An architectural overview of the constructed system is depicted in Figure 9d, showcasing the input image source, storage device, and processing node powered by an Orange Pi board running the adapted YOLOv10 model.

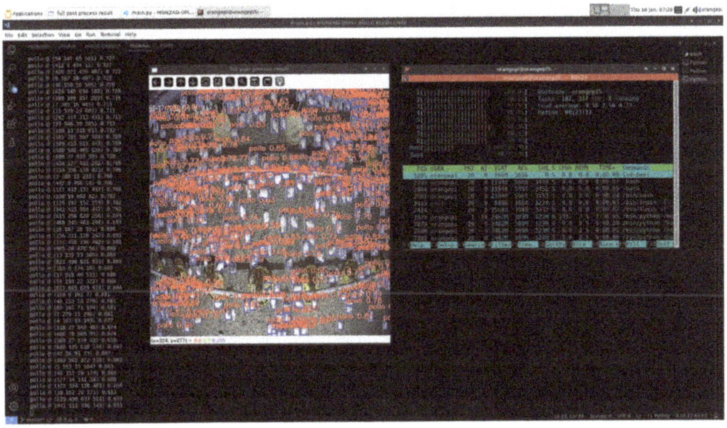

(**a**) Visualization of detected poultry during the test.

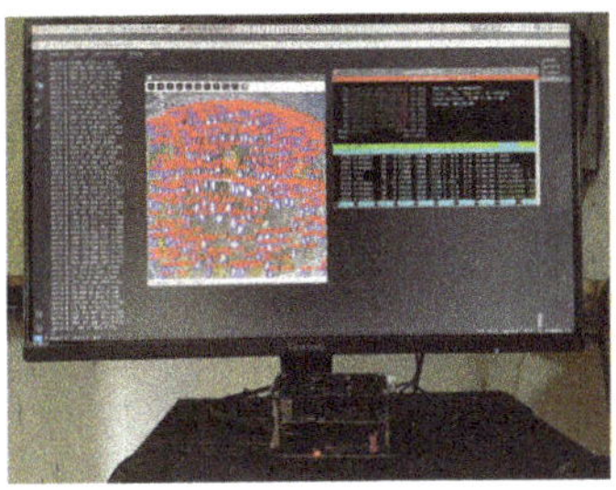

(**b**) Camera setup during real-world testing.

(**c**) System working in a real-world environment.

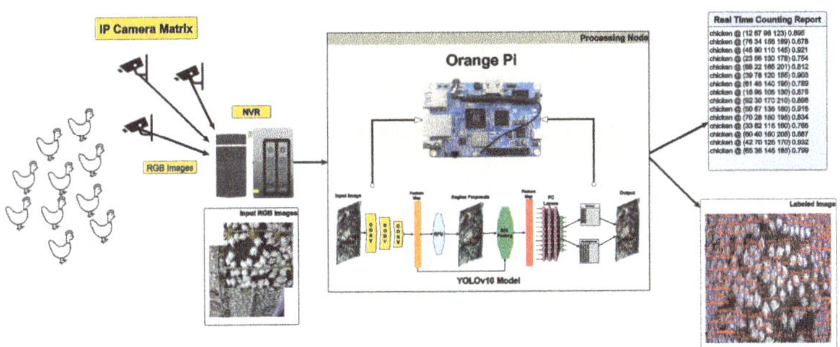

(**d**) Schematic pipeline of the system.

Figure 9. The experimental setup employed for real-world testing. This setup runs on a non-Internet-connected network.

5. Discussion

The findings of this study emphasize the significant potential of integrating advanced AI models like YOLOv10 with edge computing for poultry monitoring. The evaluation of our system highlights its high precision and recall, demonstrating its applicability for real-world farm environments. However, a thorough discussion of its relevance, as well as comparisons with existing approaches, limitations, and future implications, is necessary.

5.1. Relevance of the Results

The ACMSPT system achieved a precision of 93.1% and recall of 93.0% for poultry detection and tracking, positioning it as a highly reliable tool for automated poultry monitoring. Similar AI-driven systems have been reported in the literature. For instance, Liu et al. (2021) developed a YOLOv4-based dead chicken detection system with a precision of 95.24% [10], while in 2024, Suh et al. [27] applied YOLOv5 for classified poultry counting, with an accuracy of 89.9% [27]. Our results align with these studies, confirming the system's ability to address key challenges such as occlusions and variable lighting conditions.

The integration of edge computing plays a crucial role in real-time monitoring applications. Unlike traditional cloud-based architectures, which often introduce network latency, edge AI solutions can process data locally, reducing dependency on high-bandwidth networks [17]. This approach enhances the efficiency of automated livestock monitoring, making it a viable solution for farms in resource-limited settings [35].

5.2. Comparison with Previous Studies

Existing poultry monitoring solutions primarily rely on cloud-based architectures, which can introduce latency and require robust network infrastructure. Ren et al. (2020) emphasized the need for real-time processing in poultry automation, stating that cloud-based methods often face delays [20]. Our approach significantly improves upon these issues by achieving an inference time of less than 200 milliseconds, outperforming cloud-based alternatives which range between 500 and 1000 milliseconds [29].

Furthermore, alternative deep learning models, such as Faster R-CNN and SSD, have been used for poultry monitoring, but their computational demands are significantly higher [28]. In contrast, YOLO-based models, particularly YOLOv10, provide a balance between speed and accuracy, making them more practical for real-time edge computing applications, as shown in Table 3 where several architectures are compared with the results obtained by the ACMSPT approach.

Table 3. Comparison of poultry monitoring approaches.

Method	Latency (ms)	Accuracy (%)	Computational Demand	Deployment
Cloud-based Systems [29]	500–1000	90.5	High	Cloud-dependent
Faster R-CNN [28]	250–800	91.0	Very High	Edge/Cloud
SSD [28]	300–700	88.5	High	Edge/Cloud
YOLOv4 [10]	50–250	95.2	Moderate	Edge
YOLOv5 [27]	40–200	89.9	Moderate	Edge
YOLOv10 (ACMSPT)	**<200**	**93.1**	**Optimized**	**Edge**

5.3. Implications for Poultry Welfare and Productivity

Automated monitoring systems have been recognized for their potential to enhance poultry welfare. AI-based behavioral analysis tools can identify anomalies in movement patterns, which may indicate health concerns. Mohialdin et al. (2023) demonstrated that real-time tracking and analysis of chicken behavior contribute to early disease detection and improved flock management [54]. Furthermore, AI-driven environmental control has been shown to optimize conditions such as temperature, ventilation, and humidity, thereby reducing stress and improving growth rates [55].

Additionally, studies have explored the impact of stocking density on poultry health. Kim et al. (2025) [16] reported that increasing the available space per bird positively influenced weight gain and gut microbiome health, highlighting the importance of precision monitoring to optimize space utilization [16]. Our system's ability to provide real-time tracking supports precision agriculture strategies that enhance poultry welfare while maintaining operational efficiency.

5.4. Limitations and Future Research

While the ACMSPT system has demonstrated promising results, several limitations should be addressed. The dataset used for model training was limited to a specific farm environment. To improve generalizability, future research should expand the dataset to include different lighting conditions, poultry breeds, and environmental settings.

Another limitation concerns hardware scalability. Although the Orange Pi 5B provides a cost-effective edge computing platform, larger-scale farms may require more powerful hardware solutions, such as NVIDIA Jetson or FPGA-based implementations [17]. Future research should explore the trade-offs between computational efficiency and scalability to optimize system performance for large-scale applications.

Furthermore, while the current system focuses on tracking and counting, integrating predictive analytics for disease detection would enhance its functionality. Recent studies on gut microbiome analysis suggest that microbiome composition can serve as an early indicator of poultry health [8]. Incorporating such data sources could improve early intervention strategies and reduce disease-related losses in poultry farming.

5.5. Section Conclusion

This study contributes to AI-driven livestock management by demonstrating the feasibility of a low-latency, edge AI-based poultry monitoring system. By leveraging YOLOv10 and edge computing, ACMSPT achieves real-time poultry tracking with high precision and recall, outperforming conventional monitoring approaches. The system's advantages in speed, scalability, and operational autonomy make it a viable solution for modern poultry farming.

Future work should focus on expanding dataset diversity, integrating predictive analytics for disease detection, and optimizing hardware for large-scale deployment. With continued advancements in AI and IoT, automated livestock monitoring systems hold significant potential to revolutionize the agricultural industry.

6. Conclusions

The proposed and implemented automated monitoring system based on artificial intelligence and edge computing has proven to be an effective and efficient tool for managing chicken populations on poultry farms. Its main benefits include the ability to perform accurate real-time counts, identify movement patterns, and issue alerts for abnormal behaviors, enabling timely interventions. Additionally, the system reduces reliance on human resources and operational costs while enhancing animal welfare through constant monitoring.

In the future, the system could benefit from the integration of additional sensors to evaluate environmental parameters such as temperature and humidity, providing a more comprehensive view of farm conditions. Moreover, the implementation of more advanced AI models could improve the system's accuracy in complex environments. Exploring ways to miniaturize the hardware could also make the technology more accessible for small farms, while integrating predictive capabilities could help anticipate health or behavioral issues.

This work contributes to the advancement of technologies applied to the agribusiness sector, offering an innovative and scalable solution for automated animal monitoring on farms. By combining artificial intelligence, edge computing, and IoT, the system not only enhances operational efficiency but also establishes a sustainable and replicable model for other applications in precision agriculture. Furthermore, it promotes a transition toward more modern and responsible practices in the agricultural industry, improving sustainability and animal welfare.

Author Contributions: E.C. and M.H.-R.: These authors contributed equally to this work. E.C.: conceptualization, funding acquisition, project administration, resources, software, writing—original draft, and writing—review and editing. M.H.-R.: Conceptualization, software, data curation, formal analysis, methodology, writing—original draft, and writing—review and editing. A.M.A.-R.: resources, validation, supervision, and writing—review and editing. J.C.R.: investigation, data curation, visualization, and writing—original draft. K.B.: validation, formal analysis, and writing—review and editing. F.G.-O.: formal analysis, investigation, visualization, and writing—review and editing. All authors have read and agreed to the published version of the manuscript.

Funding: This research was funded by Project IDDS22-28 of the National Secretariat for Science, Technology, and Innovation (SENACYT), Panama. José Carlos Rangel and Edmanuel Cruz were supported by SENACYT's National Research System (SNI). Keyla Boniche received a scholarship from SENACYT's Program for the Strengthening of National Graduate Programs for a Master's Degree in Mechanical Engineering Sciences.

Institutional Review Board Statement: Not applicable.

Data Availability Statement: Data are available upon request due to restrictions.

Acknowledgments: The authors thank the IDDS Program and the National Research System (SNI) of SENACYT, as well as CEMCIT, the Veraguas Regional Center of the Universidad Tecnológica de Panamá, and Agropecuaria El Buen Pastor, for their invaluable support.

Conflicts of Interest: The authors declare that the research was conducted in the absence of any commercial or financial relationships that could be construed as a potential conflicts of interest.

References

1. Baktir, A.C.; Ozgovde, A.; Ersoy, C. How Can Edge Computing Benefit From Software-Defined Networking: A Survey, Use Cases, and Future Directions. *IEEE Commun. Surv. Tutorials* **2017**, *19*, 2359–2391. [CrossRef]
2. Katz, Y. Manufacturing an Artificial Intelligence Revolution. 2017. Available online: https://dash.harvard.edu/handle/1/37370311 (accessed on 3 March 2025).
3. Wu, Y.; Yang, Z.; Liu, Y. Internet-of-things-based multiple-sensor monitoring system for soil information diagnosis using a smartphone. *Micromachines* **2023**, *14*, 1395. [CrossRef] [PubMed]
4. Paramathma, M.K.; Kumar, I.B.; Karuppasamypandiyan, M. YOLO Based Automatic Poultry Monitoring System. In Proceedings of the 2024 3rd International Conference for Advancement in Technology (ICONAT), GOA, India, 6–8 September 2024. [CrossRef]
5. Vidya, N.L.; Meghana, M.; Ravi, P.; Kumar, N. Virtual Fencing using Yolo Framework in Agriculture Field. In Proceedings of the 2021 Third International Conference on Intelligent Communication Technologies and Virtual Mobile Networks (ICICV), Tirunelveli, India, 4–6 February 2021; pp. 441–446. [CrossRef]
6. Flora, G.D.; Sandhiya, R.; Santhiya, P.; Sindhu, G.; Sribavathaarani, K.; Ramani, U. Smart Monitoring Instrumentation Design for Poultry Farmhouse. In Proceedings of the 2022 7th International Conference on Communication and Electronics Systems (ICCES), Coimbatore, India, 22–24 June 2022; pp. 1551–1554. [CrossRef]
7. Kasparaviciene, B.; Novoslavskij, A.; Aksomaitiene, J.; Stankeviciene, J.; Kasetiene, N.; Sinkevicius, R.; Malakauskas, M. Prevalence and Antimicrobial Resistance of ESBL *E. coli* in Early Broiler Production Stage and Farm Environment in Lithuania. *Microorganisms* **2025**, *13*, 425. [CrossRef]
8. Nair, A.; Doijad, S.P.; Suryavanshi, M.V.; Dey, A.; Singh Malik, S.V.; Dutilh, B.E.; Barbuddhe, S.B. Unveiling the Kadaknath Gut Microbiome: Early Growth Phase Spatiotemporal Diversity. *Microbiol. Res.* **2025**, *16*, 54. [CrossRef]
9. Mehdizadeh, S.A.; Siriani, A.L.R.; Pereira, D.F. Optimizing Deep Learning Algorithms for Effective Chicken Tracking through Image Processing. *AgriEngineering* **2024**, *6*, 2749–2767. [CrossRef]
10. Liu, H.; Chen, C.; Tsai, Y.; Hsieh, K.; Lin, H. Identifying Images of Dead Chickens with a Chicken Removal System Integrated with a Deep Learning Algorithm. *Sensors* **2021**, *21*, 3579. [CrossRef]
11. Bist, R.B.; Bist, K.; Poudel, S.; Subedi, D.; Yang, X.; Paneru, B.; Mani, S.; Wang, D.; Chai, L. Sustainable poultry farming practices: A critical review of current strategies and future prospects. *Poult. Sci.* **2024**, *103*, 104295. [CrossRef]
12. Rao, D.R.S. Trends and challenges of poultry industry. *Int. J. Eng. Technol. Manag. Res.* **2015**, *1*, 8–13.

13. Rowe, E.; Dawkins, M.S.; Gebhardt-Henrich, S.G. A Systematic Review of Precision Livestock Farming in the Poultry Sector: Is Technology Focused on Improving Bird Welfare? *Animals* **2019**, *9*, 614. [CrossRef]
14. Fritz, S.; See, L.; Bayas, J.C.L.; Waldner, F.; Jacques, D.; Becker-Reshef, I.; Whitcraft, A.; Baruth, B.; Bonifacio, R.; Crutchfield, J.; et al. A comparison of global agricultural monitoring systems and current gaps. *Agric. Syst.* **2018**, *168*, 258–272. [CrossRef]
15. Moura, D.J.; Nääs, I.A.; Pereira, D.F.; Silva, R.B.T.R.; Camargo, G.A. Animal Welfare Concepts and Strategy for Poultry Production: A Review. *Braz. J. Poult. Sci.* **2006**, *8*, 137–148. [CrossRef]
16. Kim, E.B.; Choi, S.; Park, J.; Xuan, B. Effects of Increased Housing Space Without Altering Stocking Density on Body Weight, Stress, and Gut Microbiome in Broiler Chickens. *Animals* **2025**, *15*, 441. [CrossRef]
17. Zhang, X.; Cao, Z.; Dong, W. Overview of Edge Computing in the Agricultural Internet of Things: Key Technologies, Applications, Challenges. *IEEE Access* **2020**, *8*, 141748–141759. [CrossRef]
18. Akhtar, M.N.; Shaikh, A.J.; Khan, A.; Awais, H.; Bakar, E.A.; Othman, A.R. Smart Sensing with Edge Computing in Precision Agriculture for Soil Assessment and Heavy Metal Monitoring: A Review. *Agriculture* **2021**, *11*, 475. [CrossRef]
19. Javaid, M.; Haleem, A.; Khan, I.H.; Suman, R. Understanding the Potential Applications of Artificial Intelligence in Agriculture Sector. *Adv. Agrochem.* **2023**, *2*, 15–30. [CrossRef]
20. Ren, G.; Lin, T.; Ying, Y.; Chowdhary, G.; Ting, K. Agricultural robotics research applicable to poultry production: A review. *Comput. Electron. Agric.* **2020**, *169*, 105216. [CrossRef]
21. Gardezi, M.; Joshi, B.; Rizzo, D.M.; Ryan, M.; Prutzer, E.; Brugler, S.; Dadkhah, A. Artificial Intelligence in Farming: Challenges and Opportunities for Building Trust. *Agron. J.* **2023**, *116*, 1217–1228. [CrossRef]
22. Kaleem, A.; Hussain, S.; Aqib, M.; Cheema, M.J.M.; Saleem, S.R.; Farooq, U. Development Challenges of Fruit-Harvesting Robotic Arms: A Critical Review. *AgriEngineering* **2023**, *5*, 2216–2237. [CrossRef]
23. Khan, H.A.; Farooq, U.; Saleem, S.R.; Rehman, U.u.; Tahir, M.N.; Iqbal, T.; Cheema, M.J.M.; Aslam, M.A.; Hussain, S. Design and development of machine vision robotic arm for vegetable crops in hydroponics. *Smart Agric. Technol.* **2024**, *9*, 100628. [CrossRef]
24. Kalyani, Y.; Collier, R. A Systematic Survey on the Role of Cloud, Fog, and Edge Computing Combination in Smart Agriculture. *Sensors* **2021**, *21*, 5922. [CrossRef]
25. Redmon, J.; Divvala, S.; Girshick, R.; Farhadi, A. You Only Look Once: Unified, Real-Time Object Detection. In Proceedings of the IEEE Conference on Computer Vision and Pattern Recognition (CVPR), Las Vegas, NV, USA, 27–30 June 2016.
26. Ren, S.; He, K.; Girshick, R.; Sun, J. Faster R-CNN: Towards Real-Time Object Detection with Region Proposal Networks. *IEEE Trans. Pattern Anal. Mach. Intell.* **2017**, *39*, 1137–1149. [CrossRef] [PubMed]
27. Sun, E.; Xiao, Z.; Yuan, F.; Wang, Z.; Ma, G.; Liu, J. Method of Classified Counting of Mixed Breeding Chickens Based on YOLOV5. In Proceedings of the 2023 42nd Chinese Control Conference (CCC), Tianjin, China, 24–26 July 2023. [CrossRef]
28. Zhang, X.; He, K.; Zhao, M.; Jin, X. Object Detection Method and its Application Based on YOLOv5. In Proceedings of the 2023 2nd International Conference on Artificial Intelligence, Human-Computer Interaction and Robotics (AIHCIR), Tianjin, China, 8–10 December 2023; pp. 441–445.
29. Neethirajan, S. ChickTrack—A quantitative tracking tool for measuring chicken activity. *Measurement* **2022**, *191*, 110819. [CrossRef]
30. Hewe, T.R.A.; Mondia, K.M.J.; Empas, P.E.G. Advancing Poultry Farming Efficiency Through YOLOv5 and Image Regression-Based Broiler Weight Estimation. In Proceedings of the 2024 15th International Conference on Information and Communication Technology Convergence (ICTC), Jeju Island, Republic of Korea, 16–18 October 2024; pp. 7–12. [CrossRef]
31. Ali, W.; Ud Din, I.; Almogren, A.; Rodrigues, J.J.P.C. Poultry Health Monitoring With Advanced Imaging: Toward Next-Generation Agricultural Applications in Consumer Electronics. *IEEE Trans. Consum. Electron.* **2024**, *70*, 7147–7154. [CrossRef]
32. Chemme, K.S.; Alitappeh, R.J. An End-to-End Model for Chicken Detection in a Cluttered Environment. In Proceedings of the 2024 13th Iranian/3rd International Machine Vision and Image Processing Conference (MVIP), Tehran, Iran, 6–7 March 2024; pp. 1–7. [CrossRef]
33. Elmessery, W.M.; Gutiérrez, J.; Abd El-Wahhab, G.G.; Elkhaiat, I.A.; El-Soaly, I.S.; Alhag, S.K.; Al-Shuraym, L.A.; Akela, M.A.; Moghanm, F.S.; Abdelshafie, M.F. YOLO-Based Model for Automatic Detection of Broiler Pathological Phenomena through Visual and Thermal Images in Intensive Poultry Houses. *Agriculture* **2023**, *13*, 1527. [CrossRef]
34. Yang, X.; Bist, R.B.; Paneru, B.; Chai, L. Deep Learning Methods for Tracking the Locomotion of Individual Chickens. *Animals* **2024**, *14*, 911. [CrossRef]
35. Cruz, E.; Hidalgo-Rodriguez, M.; Acosta-Reyes, A.M.; Rangel, J.C.; Boniche, K. AI-Based Monitoring for Enhanced Poultry Flock Management. *Agriculture* **2024**, *14*, 2187. [CrossRef]
36. Subedi, K.; Sharma, S.; Paneru, P. IoT Based Smart Poultry Management System. *Int. J. Electron. Commun. Eng.* **2023**, *12*, 45–53. [CrossRef]
37. Sangeetha, M.K.; Kanthimathi, D.M.; Monisha, S.R.; Reethika, M.; Amirthabowmiya, M. Poultry Farm Control and Management System Using Wireless Sensor Networks. In Proceedings of the IEEE 2nd International Symposium on Telecommunication Technologies (ISTT), Chennai, India, 9–10 November 2014. [CrossRef]

38. Čakić, S.; Popović, T.; Krčo, S.; Nedić, D.; Babić, D. Developing Object Detection Models for Camera Applications in Smart Poultry Farms. In Proceedings of the 2022 IEEE International Conference on Omni-Layer Intelligent Systems (COINS), Barcelona, Spain, 1–3 August 2022. [CrossRef]
39. Ali, M.L.; Zhang, Z. The YOLO Framework: A Comprehensive Review of Evolution, Applications, and Benchmarks in Object Detection. *Computers* **2024**, *13*, 336. [CrossRef]
40. Alif, M.A.R.; Hussain, M. YOLOv1 to YOLOv10: A comprehensive review of YOLO variants and their application in the agricultural domain. *arXiv* **2024**, arXiv:2406.10139.
41. Carvalho, G.; Cabral, B.; Pereira, V.; Bernardino, J. Edge computing: Current trends, research challenges and future directions. *Computing* **2021**, *103*, 993–1023. [CrossRef]
42. Terven, J.R.; Cordova-Esparza, D.M. A Comprehensive Review of YOLO Architectures in Computer Vision: From YOLOv1 to YOLOv10 and YOLO-NAS. *arXiv* **2024**, arXiv:2304.00501.
43. MacPherson, J.; Voglhuber-Slavinsky, A.; Olbrisch, M.; Schöbel, P.; Dönitz, E.; Mouratiadou, I.; Helming, K. Future agricultural systems and the role of digitalization for achieving sustainability goals. *Agron. Sustain. Dev.* **2022**, *42*, 70. [CrossRef] [PubMed]
44. Neethirajan, S. Artificial Intelligence and Sensor Innovations: Enhancing Livestock Welfare with a Human-Centric Approach. *Hum.-Centric Intell. Syst.* **2023**, *4*, 77–92. [CrossRef]
45. Feng, L.; Tung, F.; Ahmed, M.O.; Bengio, Y.; Hajimirsadegh, H. Were RNNs All We Needed? *arXiv* **2024**, arXiv:2410.01201.
46. Yaseen, M. What is YOLOv10: An In-Depth Exploration of the Internal Features of the Next-Generation Object Detector. *arXiv* **2024**, arXiv:2408.15857.
47. Tzutalin. LabelImg. 2015. Available online: https://github.com/tzutalin/labelImg (accessed on 15 April 2024).
48. Gao, Y.; Xiong, Y.; Gao, X.; Jia, K.; Pan, J.; Bi, Y.; Dai, Y.; Sun, J.; Wang, M.; Wang, H. Retrieval-Augmented Generation for Large Language Models: A Survey. *arXiv* **2023**, arXiv:2312.10997.
49. Q-engineering. Qengineering/YoloV10-NPU. 2025. Available online: https://github.com/Qengineering/YoloV10-NPU (accessed on 3 March 2025).
50. Ultralytics. Despliegue de Ultralytics YOLO11 en Rockchip Para una IA de Borde Eficiente. 2025. Available online: https://www.ultralytics.com/es/blog/deploy-ultralytics-yolo11-on-rockship-for-efficient-edge-ai (accessed on 1 October 2024).
51. Ultralytics. Rockchip RKNN. 2025. Available online: https://docs.ultralytics.com/es/integrations/rockchip-rknn (accessed on 2 October 2024).
52. airockchip. airockchip/yolov10. 2024. Available online: https://github.com/airockchip/yolov10 (accessed on 2 October 2024).
53. airockchip. airockchip/rknn_model_zoo. 2025. Available online: https://github.com/airockchip/rknn_model_zoo (accessed on 2 October 2024).
54. Mohialdin, A.; Elbarrany, A.; Atia, A. Chicken Behavior Analysis for Surveillance in Poultry Farms. *Int. J. Adv. Comput. Sci. Appl.* **2023**, *14*, 935–942. [CrossRef]
55. Taleb, H.M.; Mahrose, K.; Abdel-Halim, A.A.; Kasem, H.; Ramadan, G.S.; Fouad, A.M.; Khafaga, A.F.; Khalifa, N.E.; Kamal, M.; Salem, H.M.; et al. Using Artificial Intelligence to Improve Poultry Productivity—A Review. *Ann. Anim. Sci.* **2025**, *25*, 23–33. [CrossRef]

AgriEngineering

MDPI

Article

Efficiency and Reliability of Broiler Weighing Methods in Commercial Environments: A Comparative Evaluation

Isis Mariana Dombrowsky Leal Pasian [1,*], Robson Mateus Freitas Silveira [1,2], Jessica Nacarato Reple [3], Hilton Tadeu Zarate Couto [4,†] and Iran José Oliveira da Silva [1]

1 Enrivonment Livestock Research Center, Department of Biosystems Engeneering, Luiz de Queiroz College of Agriculture, University ofSão Paulo (ESALQ/USP), Av. Pádua Dias, 11, Piracicaba 13418-900, SP, Brazil; robsonmateusfs1994@gmail.com (R.M.F.S.); iranoliveira@usp.br (I.J.O.d.S.)

2 Department of Animal Science, Luiz de Queiroz College of Agriculture, University of São Paulo (ESALQ/USP), Av. Pádua Dias, 11, Piracicaba 13418-900, SP, Brazil

3 Elanco Animal Health, Av. das Nações Unidas, 14,401, Jequitibá Tower, São Paulo 04794-000, SP, Brazil; jessica.nacarato_reple@elancoah.com

4 Department of Forest Sciences, Luiz de Queiroz College of Agriculture, University of São Paulo (ESALQ/USP), Av. Pádua Dias, 11, Piracicaba 13418-900, SP, Brazil; hilton_couto@usp.br

* Correspondence: isismari.pasian@gmail.com

† In memoriam.

Abstract: Measuring the weight of broilers is one of the most important yet labor-intensive metrics to monitor throughout a flock's development. This study aimed to comparatively assess two broiler weighing systems in a commercial production system: an automatic weighing system using a suspended platform, and a manual weighing system. Six flocks, comprising 25,000 birds each, were monitored weekly, and the weight results obtained by manual and automatic methods were compared. Up to the third week of this study, the birds were restricted to the central region of the shed, where the broiler coop was located. From the fourth week onwards, the birds were distributed into four sectors within the shed, divided by fences. Differences in weight were found between the regions of the sheds for the automatic weighing, which demonstrates that the use of an automatic scale for each division is necessary. For the manual weighing, the differences were only found in the last week of rearing, suggesting that throughout the cycle, the weighings could be performed in a single quadrant, representing the shed. Regarding the weighing method, there were statistical differences between manual and automatic weighing. The average values for automatic weighing were 1% lower than the average values for manual weighing. However, from a commercial point of view, this small difference between the methods does not impact the poultry industry. The rational use of automatic scales is recommended to optimize the monitoring of broiler chicken performance, reduce excessive handling and, consequently, minimize animal stress, promoting greater well-being.

Keywords: broilers; weighing system; weight prediction; precision animal husbandry

1. Introduction

Flock weighing is a fundamental practice in poultry farming, as it allows for the continuous monitoring of bird development. Although manual weighing is still widely used in commercial systems, it presents operational limitations. With advances in microelectronics, automated systems have been developed to optimize this process. However, the adoption of such technologies faces resistance and requires technical and scientific comparisons with traditional methods. Therefore, it is necessary to evaluate the efficiency of different weighing systems available, considering current technological advancements [1].

Manual weighing, which predominates in commercial poultry farms, is considered inefficient due to the extensive time required per bird (averaging 7 to 27 s per animal—observational data), the physical effort demanded from the staff, and the stress caused to the birds by handling [2,3]. Additionally, it is prone to reading errors and sampling bias, especially due to the tendency to capture heavier and less agile birds [4,5].

Automatic weighing consists of one or more platforms connected to an analysis and/or data transfer unit. The birds voluntarily step onto the weighing platform and data are recorded. Some scales have technology to detect and differentiate the presence of one or more birds, as well as disregarding waste deposited on the weighing platform. Automatic weighing would allow for the continuous monitoring of the weight of the flock, without placing stress on the birds or requiring a high intensity of labor in order to obtain this information. One of the questions regarding weighing sampling is whether or not it is necessary to perform it in different areas in the shed. This is true for both the manual and automatic methods.

Despite technological advances in the automation of broiler weighing, there are still limitations regarding the reliability of the data generated, especially during the final stages of the rearing period. Previous studies have reported inconsistencies between weights obtained through automatic and manual methods, with divergent results in terms of accuracy and sample representativeness [1,3,6]. While some authors have found equivalence between weighing methods, others have observed the under- or overestimation of both the weights and the number of birds assessed automatically [7–10]. These discrepancies highlight the need for a systematic evaluation of the efficiency and reliability of different weighing methods in commercial environments, considering factors that may influence the quality of the data collected throughout the production cycle. Given the previously identified issues, we hypothesize that manual weighing differs from automatic weighing, and that weight variations exist across different regions of the shed.

The objective of this study was to evaluate the existence of discrepancies between manual and automatic weighing systems with the use of platforms in real production situations. Specifically, we want to (i) compare, at each stage of flock development, the results of automatic weighing technology with manual weighing, and (ii) analyze the effect of different regions of the shed on the live weight of the birds using manual and automatic scales.

2. Materials and Methods

2.1. Characterization of the Experimental Area and Animals

This research was developed on a commercial broiler farm located in the city of Itu—SP (23°19′02.0″ S 47°23′05.2″ W). The climate of the region is CWA according to the Köppen Classification. Two production cycles were evaluated.

This study was conducted in an agro-industrial production complex consisting of 10 poultry sheds (150 m long, 12 m wide, and 2.5 m high) and a blue house model, oriented in an east–west direction. The sheds had an automatic line of feeders and nipple drinkers. The adiabatic evaporative cooling system consisted of cellulose evaporative plates, with negative pressure through the use of eight exhaust fans on the transverse walls opposite to the direction of air intake in the facility. The distance between the sheds was 15 m. Production was characterized by an average density of 14 birds/m^2, totaling 25 thousand birds housed in each shed. Two complete production cycles were evaluated, with three sheds being evaluated in each cycle, totaling six sheds, and 150,000 mixed-sex Cobb and Hubbard Flex birds were used. Feeding was automated on three feeding lines, ad libitum, according to the phases (pre-starter, starter, growth, and slaughter), and the vaccination and medication protocol was carried out in accordance with the company's guidelines.

2.2. Description of Weighing Systems

2.2.1. Automatic Weighing

Automatic weighing was performed continuously; however, for comparison between the average weight of the flock, the weighings related to the day of manual weighing were considered. The scale used for automatic weighing in this study was the BAT2 model (Veit Electronics®, Moravany, Czech Republic). It had a maximum operational capacity of 100 kg, an accuracy of 0.1%, and a resolution of 1 g. The scale was programmed so that the weights were recorded when the birds climbed up and down the weighing platform. A weight correction curve was used with the addition of 7% according to previous experiences under commercial conditions (pilot test), higher than that recommended by the supplier.

Four automatic scales were installed in each shed, distributed along the length of the facility as shown in Figure 1. This distribution was not performed so that the platforms were equidistant. In this case, it was decided to centralize them on the partitions which previously existed in the shed due to the management with migration fences adopted by the producer. The scales were positioned between the line of feeders and drinkers.

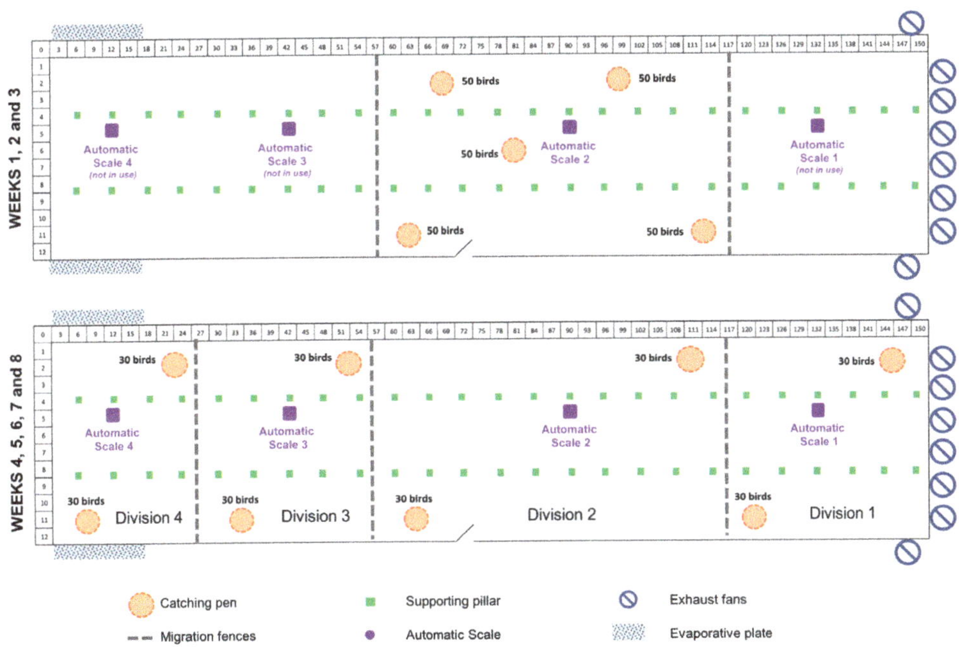

Figure 1. Positions of automatic scales and manual weighing in the stables.

It is worth noting that between 1 and 20 days, the birds remained only in Division 2, where the "chicken coop" was located. Therefore, only scale 2 of each shed was used during this period. As the flock grew, the area available for the birds increased, and consequently the number of scales that were used increased.

The location and use of the scales in the experimental area can be seen in Figure 2.

Figure 2. Scales installed in commercial sheds. (**A**) One-day-old birds on the weighing platform of the automatic scale. (**B**) Detail on the platform height (**C**). Twenty-eight-day-old birds on the weighing platform of the automatic scale, where I is the platform and II is the registration unit.

2.2.2. Manual Weighing

Manual weighings were performed weekly according to the growth of the birds, ending in the 7th week. It should be considered that the flocks were not all housed on the same day, thus presenting an age window between the flocks of each phase. Therefore, the weighings were planned to meet the demands of each flock. Since the housings were carried out on different days, in order to standardize the weighings every 7 days, there was an adjustment in the readings between the first and second week. Therefore, from the second week onwards, the weighings were performed at weekly intervals, as described in Table 1.

Table 1. Housing dates and weighing ages.

Flock	Weighing Ages
I	1; 10; 17; 24; 31; 38; 45
II	1; 10; 17; 24; 31; 38; 45
III	1; 9; 16; 23; 30; 37; 44
IV	1; 8; 15; 21; 28; 36; 42
V	1; 8; 15; 21; 28; 36; 42
VI	1; 8; 15; 21; 28; 36; 42

The BAT1 digital scale (Veit Electronics) was used for manual weighing. The maximum operational capacity was up to 50 kg and the resolution was 1 g.

According to the age group of the birds, different handling methods were adopted to adapt to their characteristics, using plates and bags for weighing as follows (Figure 3):

- Weighing plate—1st to 3rd week;
- Weighing bag—4th to 7th week.

Manual weighing always occurred in the quadrant where the automatic scale was installed. Weighing management followed the changes that commonly occur in a production system, according to the opening of the brooders. These, in turn, are opened and released according to the development of the birds. Therefore, the spatial variability of manual weighing positioning followed this management. The distribution layout of the automatic scales and manual weighing positions can be seen in Figure 1.

It should be considered that there was a weekly adjustment depending on the opening of the breeding circles and their age. The general scheme of animals sampled during the production cycle can be seen in Table 2.

Figure 3. Weighing methodology. (**A**) One-day-old bird being weighed on a suitable plate. (**B**) Twenty-eight-day-old bird being weighed in a bag.

Table 2. Sampling used according to bird age.

Age	Collection Points and Number of Birds	Total Sample per Shed
Day 1	500 animals in the pen	500
Up to the 20th day	5 points with 50 animals each	250
From the 21st day	4 quadrants × 2 points × 30 animals each	240

The decision to adopt 30 birds per measurement point was based on a previous experience (pilot test), in which all the fenced birds were initially weighed at the time of weighing. After preliminary analyses of these data (variance stabilization analysis), it was concluded that it was not necessary to collect data from all fenced birds. Therefore, 30 birds were evaluated per weighing point.

2.3. Statistical Methods

In order to study the divisions in animal performance, a completely randomized design was adopted for the division variables from the fourth week onwards. Each weighing in each division during the week was considered a repetition. First, the assumptions of data normality and homogeneity of variance were tested using the Kolmogorov–Smirnov (normality) and Box–Cox (homoscedasticity) tests. Thus, the dependent variables, weight and temperature, were transformed in order to stabilize the variance. After that, an analysis of variance (ANOVA) was performed to test the hypotheses at a probability level of 95%. The ANOVA was performed using the GLM (General Linear Models) procedure of the SAS 9.3, 2011 (Statistical Analysis System) statistical program, considering the Sum of Squares type III due to the sample imbalance for all designs. Finally, a Tukey–Kramer mean comparison test was performed, at a probability level of 95%, to determine in which treatment there is a difference between the estimated means, if the ANOVA indicated a significant difference between the treatments for each situation studied.

To represent the structure of the statistical analysis, the following equation describes the general linear model (GLM) applied in the analysis of variance:

$$Y_{ij} = \mu + \alpha_i + \varepsilon_{ij}$$

where Y_{ij} = observed value of the dependent variable (weight or temperature) in the *i*th division and *j*th repetition; μ = overall mean; α_i = fixed effect of the ith division; and ε_{ij} = random error associated with each observation, assumed to be normally distributed with mean zero and constant variance, $\varepsilon_{ij} \sim N(0, \sigma^2)$.

3. Results and Discussion

The highest frequency of weighings was recorded in the second week (2385 records), while the lowest occurred in the seventh week (718 records). A decreasing linear trend was also observed in the number of birds accessing the platform over the weeks.

3.1. Comparative Analysis of Shed Regions

There were differences ($p < 0.005$) between the partitions of the broiler sheds in all divisions from the 4th to 7th week when the birds were weighed with automatic scales. The opposite was observed for the birds weighed with manual scales, in which the weights between the partitions were similar ($p > 0.05$) in the weeks evaluated, with the exception of the last week of evaluation (Table 3).

Table 3. Average automatic and manual weight per week according to spatial variability of partitions (data from grouped phases from barn 2).

Week	Automatic				Manual			
	Partitions				Partitions			
	1	2	3	4	1	2	3	4
4	0.877 a	0.866 a	0.874 a	0.861 a	0.864 a	0.881 a	0.858 a	0.869 a
5	1.403 b	1.400 b	1.432 a	1.406 b	1.425 a	1.423 a	1.403 a	1.432 a
6	2.033 b	2.025 b	2.067 a	2.054 a	2.048 a	1.998 a	2.081 a	2.044 a
7	2.524 bc	2.507 c	2.545 ab	2.557 a	2.508 c	2.509 bc	2.637 ab	2.698 a

Equal lowercase letters correspond to statistically equal values in the same row according to the Tukey–Kramer test at a significance level of 5%, analyzed separately for automatic and manual weighing.

This result is explained by the difference in the sample size of these two methods, since manual weighing collected 60 birds in each division and automatic scales easily reached the mark of 1000 daily records. However, it is worth noting from a commercial point of view that the differences between the maximum and minimum values between the partitions are small, except in the last week, which would not have an impact on the poultry industry, although there is a statistical difference between the partitions of the sheds, especially in the automatic case.

The lower weights of the birds in division 4, both in the sheds with automatic or manual scales, can be explained by the higher air temperature in division 4 (Table 4), which possibly caused thermal stress in the animals, affecting their performance [11].

Table 4. Average air temperature in broiler houses in different partitions.

Shed	Temperature in partitions (°C)				*p*-Value
	1	2	3	4	
1	27.281 a	26.737 a	25.503 b	24.498 c	<0.0001
2	27.413 a	26.528 b	25.486 c	24.198 d	<0.0001
4	28.477 a	27.361 b	25.764 c	24.558 d	<0.0001

Equal lowercase letters correspond to statistically equal values in the same row according to the Tukey–Kramer test at a 5% significance level.

This thermal variability in negative pressure ventilation sheds is quite common. Nascimento et al. [12] found a thermal difference between quadrants of negative pressure ventilation poultry houses, i.e., spatial thermal heterogeneity. Miragliotta et al. [13] also observed high spatial thermal variability in the shed, similar to what was found in this study. The authors Menegali et al. [14] and Damasceno et al. [15] commented on the inefficiency and thermal heterogeneity throughout the negative pressure ventilation system sheds for broilers. Even though partitions 4 and 3 had the lowest house temperatures compared to the others, it is worth noting that these were outside the ideal ranges (22 °C) [15]. When broilers are within the thermoneutral zone, they present higher production rates (weight gain, feed conversion, etc.) [16–18].

In general, based on the data presented in Table 3, it can be concluded that there were no statistical differences in the positioning of the dividers in the manual weighing, except in the 7th week. On the other hand, in the automatic weighings there were differences in all dividers and in all weeks, evidencing that this weighing method is more sensitive to identify common weight variability across the shed. Therefore, it is suggested that one automatic scale is not sufficiently representative of a flock and that platforms should also be placed in the different quadrants of a shed.

3.2. Comparison Between Manual and Automatic Weighing Results

The comparison between the weights obtained by the manual and automatic methods was performed at the ages at which the manual weighings occurred, as described in Table 1. For all ages, with the exception of weights in week 5, there was a statistical difference between the weights obtained by automatic and manual weighing. The results are described in Table 5.

Table 5. Average weights (kg) obtained by manual and automatic weighing.

Age (Weeks)	Weighing Type (kg)		*p*-Value	Weighing Difference (kg) (PA-PM)	Variation (%)
	Automatic	Manual			
1	0.044	0.040	$p < 0.05$	+0.004	+10
2	0.169	0.174	$p < 0.05$	−0.005	−3
3	0.447	0.454	$p < 0.05$	−0.007	−2
4	0.880	0.867	$p < 0.05$	+0.013	+1
5	1.419	1.419	NS	0.000	0
6	2.054	2.072	$p < 0.05$	−0.019	−1
7	2.531	2.582	$p < 0.05$	−0.051	−2
Average	1.087	1.078	$p < 0.05$	−0.009	−1

The largest discrepancy occurred in the 1st week of evaluation (+10%), while in the 5th week, the measurements coincided (0%). This variation can be attributed to the lower accuracy of manual scales when weighing young animals, whose reduced body weight amplifies small absolute differences in relative terms. Furthermore, manual handling is more prone to operational errors, such as animal instability, inaccurate readings, or inadequate calibration. The progressive adaptation of the animals to the automatic scale may have also contributed to the reduction in variations over time. The convergence of values in the following weeks reinforces the reliability of the automatic scales, especially after the initial adaptation phase, with the potential to enhance the accuracy of zootechnical data, minimize animal stress, and streamline management.

These results are in agreement with Newberry et al. [2] and Blokhuis et al. [7], who concluded that the average weight obtained automatically was lower than that obtained manually. However, Schomburg et al. [8] compared automatic and manual weighing, observing that automatic weighing overestimated the number of animals due to occasional problems with system recalibration, which resulted in measured weights greater than the actual weights. However, other authors found results contrary to this work, such as

Turner et al. [1,6] and Doyle and Leeson [3], who found no statistical differences between automatic and manual weighing. These research results should be considered in the way in which they were performed, whether on laboratory or production scales, and especially in the particularity of the equipment evaluated. However, some researchers obtained different results for manual and automatic weighing.

Vranken et al. [19], when comparing automatic weighing with manual weighing, found reductions of 1%; 14%; 4%; −6%; −2%, and −14% for the 2nd, 3rd, 4th, 5th, 6th, and 7th weeks, respectively. The authors did not describe whether or not they used any correction curves. These authors sought to use image analysis in conjunction with the automatic scale to create an algorithm that would help provide a better estimate of flock weight and obtained an error of 5% more than manual weighing in the 7th week.

4. Conclusions

Manual and automatic weighings showed slight differences in the weighing of birds. Although there were statistical differences between sections of the barn in weeks 4, 5, and 6, these do not compromise management decisions, allowing for the recommendation of using the automatic scale in a representative quadrant during these phases. However, in the final week of production, when weight precision is crucial for commercial and slaughter decisions, it is essential to have a scale in each quadrant, regardless of the weighing method.

Thus, the rational use of automatic scales can optimize the monitoring of broiler performance, reduce excessive handling, and consequently minimize animal stress, promoting greater welfare. Additionally, standardizing the use of scales at strategic points in the barn contributes to the technological efficiency of the production system, combining data accuracy with reduced operational costs and improved decision-making.

Author Contributions: I.M.D.L.P.—Data collection, Writing—original draft, Conceptualization, Data curation, Methodology, Writing—original draft, Validation; R.M.F.S.—Writing—original draft, Validation; J.N.R.—Data collection, Conceptualization; H.T.Z.C.—Data analysis; I.J.O.d.S.—Data collection, Writing—original draft, Conceptualization, Data curation, Methodology, Writing—original draft, Validation. All authors have read and agreed to the published version of the manuscript.

Funding: The authors declare that this study received funding from Elanco Animal Health. The funder had the following involvement with the study: material acquisition.

Institutional Review Board Statement: The experiment was approved by the Ethics Committee for the Use of Animals (CEUA) of ESALQ (Protocol 2015-12).

Informed Consent Statement: No applicable.

Data Availability Statement: The original contributions presented in this study are included in the article. Further inquiries can be directed to the corresponding author(s).

Conflicts of Interest: Author Pasian is a former employee of Elanco Animal Health. Author Reple works for Elanco Animal Health. The remaining authors declare that the research was conducted in the absence of any commercial or financial relationships that could be construed as a potential conflict of interest.

References

1. Turner, M.J.B.; Gurney, P.; Belyavin, C.G. Automatic weighing of layer-replacement pullets housed on litter or in cages. *Br. Poult. Sci.* **1983**, *24*, 33–45. [CrossRef]
2. Newberry, R.C.; Hunt, J.R.; Garriner, E.E. Behaviour of roaster chickens towards an automatic weighing perch. *Br. Poult. Sci.* **1985**, *26*, 229–237. [CrossRef]
3. Doyle, I.; Leeson, S. Automatic weighing of poultry reared on a litter floor. *Can. J. Anim. Sci.* **1989**, *69*, 1075–1081. [CrossRef]
4. Feighner, S.D.; Godowsky, E.F.; Miller, B.M. Portable microcomputer-based weighing systems: Applications in poultry science. *Poult. Sci.* **1986**, *65*, 868–873. [CrossRef]

5. Meijerhof, R. Automatic weighing system gives advantages. *Pluimveehouderij* **1989**, *19*, 12–13.
6. Turner, M.J.B.; Gurney, P.; Crowther, V.S.W.; Sharp, J.W. An automatic weighing system for poultry. *J. Agric. Eng. Res.* **1984**, *29*, 17–24. [CrossRef]
7. Blokhuis, H.J.; van der Haar, J.W.; Fuchs, J.M.M. Do weighing figures represent the flock average? *Poult. Misset Int.* **1988**, *4*, 17–19.
8. Schomburg, H.; Malchow, J.; Sanders, O.; Knöll, J.; Schrader, L. Elevated platforms with integrated weighing beams allow automatic monitoring of usage and activity in broiler chickens. *Smart Agric. Technol.* **2023**, *3*, 100095. [CrossRef]
9. Chedad, A.; Aerts, J.M.; Vrankren, E.; Lippens, M.; Zoons, J.; Berckmans, D. Do heavy broiler chickens visit automatic weighing system less than lighter birds? *Br. Poult. Sci.* **2003**, *44*, 663–668. [CrossRef] [PubMed]
10. Hughes, B.O.; Elson, H.A. The use of perches by broilers in floor pens. *Br. Poult. Sci.* **1977**, *18*, 715–722. [CrossRef]
11. Sakamoto, K.S.; Silveira, R.M.F.; Benincasa, N.C.; Contreras Castillo, C.J.; Lobos, C.M.V.; Silva, I.J.O.d. Tunnel-Ventilated Sheds with Negative Pressure Reduce Thermal Stress and Improve the Meat Quality of Broilers. *Animals* **2024**, *14*, 2017. [CrossRef] [PubMed]
12. Nascimento, G.R.; Nääs, I.A.; Baracho, M.S.; Pereira, D.F.; Neves, D.P. Termografia infravermelho na estimativa de conforto térmico de frangos de corte. *Rev. Bras. Eng. Agric. Ambient.* **2014**, *18*, 658–663. [CrossRef]
13. Miragliotta, M.Y.; Nääs, I.A.; Manzione, R.L.; Nascimento, F.F. Spatial analysis of stress conditions inside broiler house under tunnel ventilation. *Sci. Agric.* **2006**, *63*, 426–432. [CrossRef]
14. Menegali, I.; Tinôco, I.F.F.; Baêta, F.C.; Cecon, P.R.; Guimarães, M.C.C.; Cordeiro, M.B. Ambiente térmico e concentração de gases em instalações para frangos de corte no período de aquecimento. *Rev. Bras. Eng. Agric. Ambient.* **2009**, *13*, 984–990. [CrossRef]
15. Damasceno, F.A.; Yanagi Junior, T.; Lima, R.R.; Gomes, R.C.C.; Moraes, S.R.P. Avaliação do bem-estar de frangos de corte em dois galpões comerciais climatizados. *Cienc. Agrotec.* **2010**, *34*, 1031–1038. [CrossRef]
16. Nascimento, S.T.; Silva, I.J.O.; Maia, A.S.C.; Castro, A.C.; Vieira, F.M.C. Mean surface temperature prediction models for broiler chickens-a study of sensible heat flow. *Int. J. Biometeorol.* **2013**, *58*, 195–201. [CrossRef] [PubMed]
17. Furtado, D.A.; Azevedo, P.V.; Tinôco, I.F.F. Análise do conforto térmico em galpões avícolas com diferentes sistemas de acondicionamento. *Rev. Bras. Eng. Agric. Ambient.* **2003**, *7*, 559–564. [CrossRef]
18. Nazareno, A.C.; Pandorfi, H.; Almeida, G.L.P.; Giongo, P.R.; Pedrosa, E.M.R.; Cristiane Guiselini, C. Avaliação do conforto térmico e desempenho de frangos de corte sob regime de criação diferenciado. *Rev. Bras. Eng. Agric. Ambient.* **2009**, *13*, 13–22. [CrossRef]
19. Vranken, E.; Chedad, A.; Aerts, J.M.; Berckmans, D. Improving theaccuracy of automatic broiler weighing by imag analysis. In Proceedings of the 2nd European Conference of Precision Livestock Farming, Proceedings, Wageningen, The Netherlands, 6–8 July 2009; pp. 265–271.

AgriEngineering

MDPI

Article

Design and Prototyping of a Robotic Structure for Poultry Farming

Glauber da Rocha Balthazar [1,2], Robson Mateus Freitas Silveira [1,*] and Iran José Oliveira da Silva [1]

[1] Ambience Research Center, Department of Biosystems Engineering, "Luiz de Queiroz" College of Agriculture, University of São Paulo (ESALQ/USP), Av. Pádua Dias, 11-São Dimas, Piracicaba 13418-900, SP, Brazil; glauber.balthazar@usp.br (G.d.R.B.); iranoliveira@usp.br (I.J.O.d.S.)

[2] Federal Institute of Education, Science and Technology of São Paulo (IFSP) R. Heitor Lacerda Guedes, 1000—Cidade Satélite Íris, Campinas13059-581, SP, Brazil

* Correspondence: robsonsilveira@usp.br

Abstract: The identification and prediction of losses, along with environmental and behavioral analyses and animal welfare monitoring, are key drivers for the use of technologies in poultry farming which help characterize the productive environment. Among these technologies, robotics emerges as a facilitator as it provides space for the use of several computing tools for capture, analysis and prediction. This study presents the full methodology for building a robot (so called *RobôFrango*) to its application in poultry farming. The construction method was based on evolutionary prototyping that allowed knowing and testing each physical component (electronic and mechanical) for assembling the robotic structure. This approach made it possible to identify the most suitable components for the broiler production system. The results presented motors, wheels, chassis, batteries and sensors that proved to be the most adaptable to the adversities existing in poultry farms. Validation of the final constructed structure was carried out through practical execution of the robot, seeking to understand how each component behaved in a commercial broiler aviary. It was concluded that it was possible to identify the best electronic and physical equipment for building a robotic prototype to work in poultry farms, and that a final product was generated.

Keywords: robotics in poultry farming; sensors and actuators in poultry farming; RobôFrango; robot precision livestock farming; robot poultry animal welfare

1. Introduction

Poultry farming has undergone severe transformations in recent decades due to technological innovations promoted by Precision Animal Science [1,2]. Concerns such as loss reduction, environmental analysis, behavioral analysis and animal welfare are factors that drive the use of technologies such as Big Data [3], Artificial Intelligence [4–7] and robotics [8,9] as solutions for identifying technological bottlenecks in production rates. This allows the understanding of how some production gaps can be filled through the combination of electronic devices with intelligent technologies that generate analytical and predictive models on animal production points that are not easy to analyze.

Robotics enters this context as a tool that can be used both to monitor and to act in so-called "smart farms". In performance, repetitive tasks (those that require human presence in unhealthy situations/hazardous) can be automated by a mechanical and electronic device that executes such tasks according to a previously programmed result. Current research presents several robots that have the ability to collect eggs from the floor of the aviary (Figure 1A) [10,11], turn over the litter (Figure 1B), sanitize the environment and promote the movement of birds (Figure 1C) [12–14]. In monitoring, high-precision environmental sensors are used to evaluate thermal and environmental conditions that are fundamentally required for the physiological development of animals and to promote very good welfare [4,15,16], as well as to activate equipment.

(A) (B) (C)

Figure 1. Robots used in poultry farming: PoultryBot (**A**); OctopusRobot (**B**); and Spoutnic Nav (**C**). Source: [17–19].

In poultry farming, environmental monitoring relies on two main types of equipment: stationary machines and mobile robots. Fixed-location monitoring systems use sensors and cameras to provide continuous surveillance of specific areas, such as entrances, perimeters, and critical points. These systems often integrate with alarms, ensuring consistent security coverage. Mobile monitoring robots, on the other hand, adaptively navigate different environments, making them better suited for large or dynamic spaces. Their flexibility allows for route adjustments and task reprogramming. However, their mobility increases maintenance needs and energy consumption, which can affect operational efficiency. Both solutions benefit from AI and machine learning, enabling predictive behavior analysis and real-time automation to enhance monitoring precision and adaptability [20–22].

Navigating a robot in an aviary presents significant challenges due to the dynamic and unpredictable nature of the animals' environment. To mitigate disturbances, specialized autonomous movement algorithms are required to ensure smooth and non-intrusive path planning. Some systems utilize proprietary Directed Coverage Path Planning (DCPP) algorithms [23], while others adopt open-source solutions like A* (A-star) [24]. Most implementations integrate hardware, such as LiDAR sensors, with software, including computer vision, to develop proprietary autonomous navigation systems [25]. Effective robotic navigation in such dynamic settings thus depends on scenario-specific algorithm models, leveraging integrated technologies for optimal performance [20,26,27].

This study presents the hypothesis that it is possible to use rapid prototyping mechanical and electronic components to develop a mobile robot to apply in poultry farming. The objective of the study was to develop and validate a robotic prototype (*RobôFrango*) for poultry farming, focusing on selecting and testing the most suitable components for broiler production systems. The specific objectives were (a) to design and implement a robotic structure (so-called *RobôFrango*) capable of supporting electronic components for movement and data collection; (b) to develop and implement an algorithmic structure in a compatible programming language capable of performing the locomotion and actuation of the robotic structure in a commercial broiler poultry farm; and (c) to carry out the execution of this robotic structure in a commercial poultry farm to evaluate its efficiency in movement.

2. Materials and Methods

The present study follows a development-based research approach focused on technological evolution and iterative prototyping. Therefore, the structure of the manuscript intentionally adopts a chronological narrative, detailing the evolution of the robotic prototypes from initial design stages to the final solution. This structure is commonly employed in engineering and applied robotics studies [28,29] where the primary objective is the step-wise resolution of technical challenges rather than a conventional experimental hypothesis testing framework. This narrative approach allows the reader to follow the decision-making process, the identification of technological bottlenecks, and the improvements implemented at each iteration, which are fundamental outcomes in robotics development studies.

The construction technique was adapted into the Scrum software 2020 development methodology [30]. It was proposed to prepare a "product backlog" for the project's functionalities and then "sprint backlogs" were defined with a maximum execution time of 3–5 weeks, totaling a perspective of 15 months of development. The construction was divided into 3 "sprints" (Figure 2).

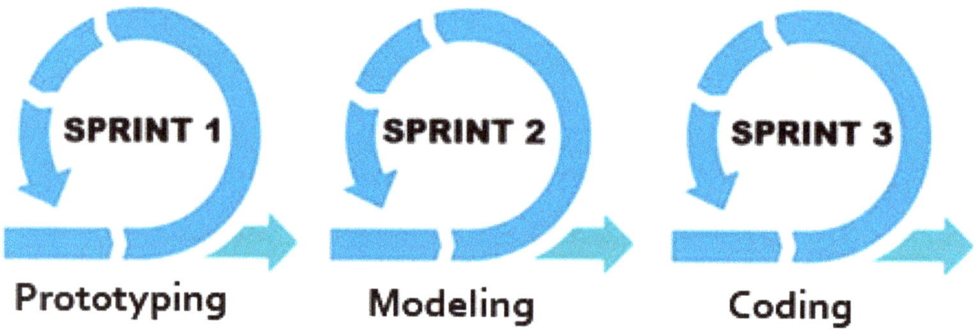

Figure 2. Constructive iterations of the robotic structure.

The details of the "sprints" are presented below:

1. **Sprint 1 (evolutionary prototypes):** identification of the most suitable electronic components for the project and their behavior in a commercial broiler aviary. This "sprint" generated several prototypes of robotic structures that were analyzed on different aspects (locomotion, load capacity, data reading, etc.); the results were several qualitative perspectives on the use of components until the adaptation of the most suitable ones for building the model and subsequent final product.
2. **Sprint 2 (3D modeling and physical implementation):** Design of the structural dimensioning of the robotic composition was carried out through modeling using a CAD (Computer-Aided Design) tool. The objective was to develop a virtual prototype that could be validated in a reduced scale format and dimensions in order to accommodate all components and their distribution to harmonize with the robot's center of gravity; the final model was built in real size using structuring materials.
3. **Sprint 3 (coding and field experiment):** implementation of an algorithm in a programming language that allowed (a) manipulating the robot's locomotion motor structure and (b) moving it around the aviary to analyze the electronic and mechanical components.

It is important to clarify that this study is focused on the technological development process of a robotic platform for poultry farming using an evolutionary prototyping approach. The objective was not to perform a statistical comparison between prototypes or to quantify the aging of mechanical and electronic components, such as wheel degradation or chemical wear. Each prototype iteration served as an engineering step to identify and solve specific design challenges related to locomotion, stability, battery autonomy, and structural adequacy under poultry house conditions. Therefore, detailed statistical analyses regarding component lifespan, debris accumulation, or physical–chemical aging were beyond the scope of this study but are acknowledged as relevant topics for future research.

It is important to note that the field tests were conducted using a single functional prototype, as it is the standard in engineering design research. To ensure repeatability and reliability of the locomotion performance assessment, three independent movement trials were performed for each motor configuration, both under unloaded and loaded conditions (with a 2.5 kg additional weight simulating the environmental sensor module). This approach is consistent with robotics validation protocols in agricultural applications [30,31], where replication occurs at the level of test runs rather than multiple physical prototypes.

This study was submitted, analyzed and approved by the Ethics Committee on the Use of Animals of the Luiz de Queiroz College of Agriculture of the University of São Paulo under protocol number 7364090322. To carry out this project, a partnership was established with a broiler chicken producer of the Cobb lineage, located in the city of Itatiba, in the state of São Paulo (latitude: −23.0068; longitude: −46.8387; 23°0′24″ S, 46°50′19″ W; altitude: 771 m). The region is characterized by a humid subtropical climate, classified as Cfa according to the Köppen–Geiger system. The aviary used was a negative pressure type, measuring 135 m in length, 14 m in width, and 3 m in ceiling height. It was equipped with evaporative cooling pads, exhaust fans, and an automatic electronic system for monitoring relative humidity and dry air temperature (inoBram brand, model SMAAI 3 PE, Pato Branco, Brazil).

3. Results and Discussion

3.1. Evolutionary Prototypes

The environment of poultry housing is too aggressive of a place for the presence of electronic components and circuits. Levels of gases (such as ammonia and carbon dioxide), dust, relative humidity and air temperature are always in fluctuation, which can affect the useful life of the chemical and physical components that make up electronic equipment. Furthermore, the floor (chicken litter) is another factor that makes the presence of robots difficult due to several factors; for example, its own material can make it difficult to move the wheels and the animals' debris (droppings: feces and urine) can accumulate in the couplers, thus increasing drag and causing corrosion. Therefore, the results of preliminary studies of different types of robotic architectures, motors, batteries, microcontrollers, etc., are presented to determine which ones would be best suited for this environment.

3.1.1. First Prototype

The first prototype was built on an aluminum structure, with a Wemos D1 microcontroller, four 6 v DC motors, 65 mm × 25 mm wheels, an L298N H bridge and three 1500 mAh 18,650 batteries (Figure 3).

(A) (B) (C)

Figure 3. First prototype built: (**A**): top left view; (**B**): tests in the poultry shed; (**C**): top right view.

This prototype was tested on a commercial poultry farm and presented the following results: the wheels were smooth and accumulated a lot of litter, in addition to being too small to elevate the structure and maintain enough free space to avoid scraping the prototype's floor; the motors were very weak and were unable to drive the structure (which was light, 745 g) in addition to not being able to make the robot perform curves (even braking the left wheels to turn to the right and vice versa); however, the H bridge was satisfactory for motor control (driver). The three 1500 mAh model 18,650 batteries had enough charge to carry out the tests and were later recharged; the aluminum structure did not guarantee adequate insulation, generating, from time to time, short circuits that shut down the entire structure; the microcontroller proved to be satisfactory, managing to support a program in C++ language that directed the robot. Finally, its size (length 25 cm × width 15 cm × height 12 cm) was not enough to be around the animals in the final weeks of the breeding cycle and it was unable to carry an environmental analysis module.

3.1.2. Second Prototype

The attempt to purchase a ready-made chassis (or structure) was made in the second prototyping (Figure 4). However, the results were not satisfactory: a resistant plastic structure was acquired with the advantage of maintaining an excellent free space above the ground (8 cm) but its dimensions (length 25 cm × width 20 cm × height 10 cm) were not sufficient to load ambience sensors. Furthermore, it only provided support for the same direct current (DC) motors and wheels as the previous robot and still did not have Battery Management System (BMS) protection despite being able to support some ambience sensors; the Wemos D1 proved to be incompatible as it has only one analog port and there are several sensors that transmit data through analog ports, requiring the insertion of another component (multiplexer) to guarantee this communication in addition to its size, which took up a lot of space in the structure. Finally, the motorization did not have enough strength to turn the robot.

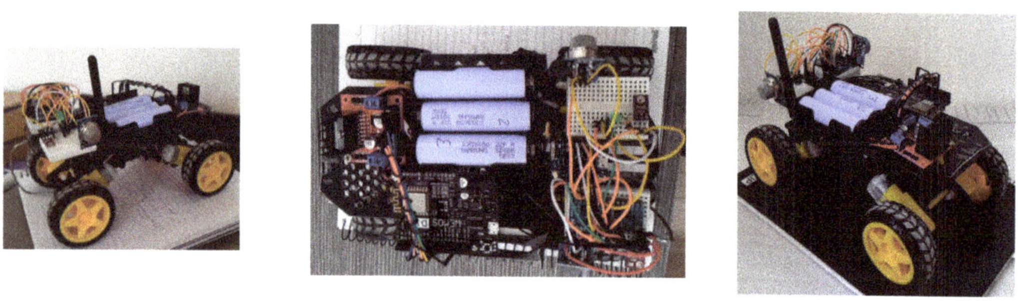

(A) (B) (C)

Figure 4. Second prototype built: (**A**): left-side view; (**B**): top view; (**C**): front view.

The fragility of the wheels led to studies that determined the best type to be used on chicken litter (specifically sawdust and peanut shells). Five types of wheels were analyzed and the fitting hole between the wheel and the motor was not considered (there are couplers that serve as a connection between the wheel and the motor shaft) (Table 1). The manufacturer was disregarded because this type of equipment generally does not have labeling.

Characteristic factors of the bedding (such as compaction, hardening, plastering, excess excrement, etc.) were analyzed qualitatively using the technical specifications of the wheels. Therefore, the five models used in this research were analyzed on two aspects: grip and height (free space) from the floor. The adhesion factor was chosen due to the different characteristics that the bedding presents throughout its useful life; the height factor was chosen due to the possibility of keeping the robot chassis as high as possible from the chicken litter, avoiding drag. For grip, all wheels that had smooth rubber or discreet trim (wheel 34 mm, wheel 68 mm and wheel 85 mm model B) were discarded as they tended to slip and accumulate debris easily. For height (free space), the 130 mm wheel and the 85 mm model A wheel proved to be compatible with the presence of both deformities on the bed surface (elevations and depressions) and cobbled parts. Finally, considering that the bedding can appear as compact, stony, plastered, muddy or soft (depending on its time of use and level of conservation), it was concluded that both the 85 mm and 130 mm wheels presented a good distance of free space for the robotic structure in relation to the ground, in addition to having sufficient grip to detach the robotic structure as they were rubberized with claws. However, the 130 mm wheel had a larger free span (65 mm) than the 85 mm wheel (42 mm) due to the robot's chassis design and the later addition of a bed humidity sensor to the structure.

Table 1. Wheels used in the research.

Wheel Model/Diameter	Tire Width	External Coating
34 mm	6.5 mm	Rubberized with friezes
68 mm	28 mm	Rubberized with discreet friezes
85 mm model A	38 mm	Rubberized with claws
5 mm model B	38 mm	Rubberized with discreet friezes
Wheel 130 mm	42 mm	Rubberized with claws

3.1.3. Third Prototype

The third prototype (Figure 5) was based on the Ackerman architecture, thus having a chassis and front axle capable of steering. In this model, the Wemos D1 mini microcontroller was chosen because it is small in size and thus created space for the presence of the environmental sensor module. The dimensions of this prototype were 55 cm (length) × 15 cm (width) × 10 cm (height).

(A) (B) (C)

Figure 5. Third prototype built: (**A**): top view; (**B**): top view with the environmental sensor prototype; (**C**): in field test view.

The steering of the wheels was enough to make the robot move in several directions, and the size made it possible to accommodate the prototype of the environmental sensor; however, the Wemos Mini has a very small number of digital ports and only one analog one, and this model still relies on very low wheels and weak DC motors. The batteries did not have a protection circuit (Battery Management System) which resulted in their malfunction (very rapid discharges), especially when their life was at the end, but their lifespan was considered good (around 1 h uninterrupted). Its movement in the aviary was satisfactory; however, it could not support weight above 1 kg (e.g., if an animal climbed onto this prototype it would not have the strength to continue its movement).

The batteries used in this third prototype were analyzed, focusing on three types: LiPO, Li-Ion and sealed batteries (Valve-Regulated Lead–Acid—VRLA). Rather than individual cells, the analysis considered battery packs designed to provide higher energy storage capacity, with protection circuits (BMS) and voltage variations between 6 v and 16 v (minimum and maximum limits supported by the Wemos Mega R3 ESP8266/ATmega2560). Table 2 presents the models evaluated.

Table 2. Batteries analyzed for the project.

Type/Model	Description
Battery Lipo 5000 80 c Uruav 11.1 v	Battery with capacity of 5000 mAh, voltage of 11.1 v, high discharge rate and weight of 383 g
Battery Lipo 11.1 v 6500 mah 45 c Uruav	Battery with capacity of 6500 mAh, voltage of 11.1 v, medium discharge rate and weight of 413 g
Battery VRLA 12 v 40.0 AH M6 UP12400 RT 06C043	VRLA type sealed battery with 40 Ah of capacity, 12 v of voltage, high discharge rate and weight of 13.5 kg
Battery Li-Ion JWS 6 × 18,650 rechargeable	18,650 battery pack (6 units), voltage of 8.4 v, capacity of 28,800 mAh weight of 285 g
Battery Li-Ion LiitoKala 12 × 18,650 rechargeable	18,650 battery pack (6 units), 12 v of voltage, 50,000 mAh of capacity, weight of 212 g

The two Lipo batteries presented voltages compatible with what was needed in the project, but they have a low charge capacity. According to the study carried out by [31] on the influence of temperature on robotic structures, the reference temperature for this type of battery is 20 °C and variations above or below 10 °C reduce its usability by around 8 to 25% (permanent). Therefore, as these batteries already have a low charge capacity and the aviary, over several weeks of breeding phases, has an ambient temperature above 30 °C, it was concluded that the temperature of the aviary would reduce the useful life of the battery and, therefore, if it had a low charging capacity, it would have a reduced useful life. Therefore, this type of battery was qualified as unsuitable for the project. The VRLA-type battery has a large charging capacity (40 Ah) and voltage compatible with the project; however, an excess weight of 13.5 kg will cause excess weight to the robotic structure, thus requiring motors with higher starting torque, due to the weight of the battery plus the robotic structure itself. This characteristic disqualified this battery, also making it incompatible with the project. The JWS Li-Ion battery pack featured a low voltage (8.4 v), a high charging capacity (28 Ah) and an adequate weight (285 g). Finally, LiitoKala's Li-Ion battery pack had an adequate voltage (12 v), a good charging capacity (10 Ah) and also an adequate weight (412 g).

Based on these considerations, it was concluded that the JWS Li-Ion battery was compatible with the project's need for powering the sensor circuit as it has a high charging capacity and voltage (8.4 v) equivalent to the Wemos R3 AtMega, but not compatible with 12 v DC motors. The Li-Ion battery had a compatible voltage (12 v) with 12 v DC motors and a good charging capacity. Thus, it was understood that the sensor circuit should be implemented separately from the movement circuit, ensuring compatible voltages with the components of each circuit and increasing the operating time capacity of the robotic structure. Therefore, two circuits are presented in the system (movement and sensing) with individual supplies. This corroborates what is presented in the PoultryBot power system, as according to [30], a 12 v battery set was made available for the robot's obstacle sensing system, with another 24 v battery for the motorization system. This was due to the fact that 24 v motors were adopted in the PoultryBot, which made the power supply system for the motorization mechanism and the sensing mechanism incompatible. In this project, voltage compatibility was guaranteed between all sensors and actuators, enabling the use of a single battery model with the same voltage.

3.1.4. Fourth Prototype

The fourth prototype (Figure 6) presented satisfactory results in terms of movement and, therefore, was completed with a cover structure (dimensions: length 45 cm × width 45 cm and height 12 cm). At this point, the environmental sensor module was coupled, and it was possible to carry out environmental analysis tests. However, its small wheels (prone to accumulating dirt and often jamming), limited weight capacity (unable to move if an animal climbed on it), low ground clearance (which caused frequent scraping on uneven surfaces), and excessive width made it unsuitable for the environment, mainly because it was unable to fend off animals (this was suggested by escape distance studies).

(A) **(B)**

Figure 6. Fourth prototype built: (**A**): top view with the environmental sensor prototype; (**B**): in field test view.

The lack of ability to carry weight led to the study of different types of motors. The motorization of the robotic structure was analyzed, considering the need for torque (starting and movement) and speed to move the chassis and its structure depending on the particularities existing in the chicken litter. Thus, the following were analyzed: the speed of the motors based on the number of revolutions per minute (rpm), the movement capacity from inertia through starting torque (post-inertia displacement), supply voltage, dimensions, and weight. Table 3 presents the four DC motors used and their characteristics.

Table 3. DC motors used in the research.

Model (Dimensions and Weight)	Voltage	Rotations	Torque (Starting and Moving)
Dual Shaft DC Motor (length: 70 mm; height: 22.5 mm; width: 37 mm; weight: 28 g)	3–6 v	80 rpm	0.35 kgf/cm (3 v) and 0.80 kgf/cm (6 v)
N20 with single axle (length: 36 mm; height: 12 mm; width: 12 mm; weight: ~15 g)	6 v	300 rpm	1.7 kgf/cm (6 v) and 0.60 kgf/cm (6 v)
DC Motor 12 v 25ga 370 (length: 55 mm; height: 23 mm; width 23 mm; weight 32 g)	6–12 v	26 rpm	9 kgf/cm and 25 kgf/cm
Dc Motor 12 v 25ga 370 (length: 60 mm; height: 25 mm; width 25 mm; weight 42 g)	6–12 v	12 rpm	12 kgf/cm and 25 kgf/cm

According to [30], the PoultryBot robot was built with three pneumatic wheels controlled by two Roboteq AX3300 controllers. This made it more complex compared to the mechanism with four motors and just one controller adopted in the present project because for the PoultryBot it was necessary to calculate the speed per individual wheel, while in this project the algorithm used in the microcontroller performed the same speed on all four wheels due to the L298n H-bridge maintaining a single speed control for all wheels.

3.1.5. Fifth Prototype

The fifth prototype was created to analyze the movement and charging capacity using these motors (Figure 7), which was assessed in a movement test. This prototype was based on the Pioneer 3-DX. However, four 12 v motors and 85 mm wheels with mudguards were included. The microcontroller was changed to the Wemos Wifi R3 ATMega2560 + ESP8266 as it has a greater number of digital and analog ports, as well as great memory and processing. Its dimensions were 45 cm in length × 35 cm in width × 41 cm in height.

(A) (B)

Figure 7. Fifth prototype built: (**A**): bottom view; (**B**): side view.

The modeling considered the spatial distribution of components in the robot chassis through the correct distribution of weights and spacing between components, ensuring the stability of the robot with analysis at the center of gravity. The motors were controlled by an L298n H-bridge driver and the Wemos Wifi ESP8266 + AtMega328p microcontroller and powered by a 12 v direct current power source. An electronic circuit diagram was designed to understand the connection scheme between the components (Figure 8).

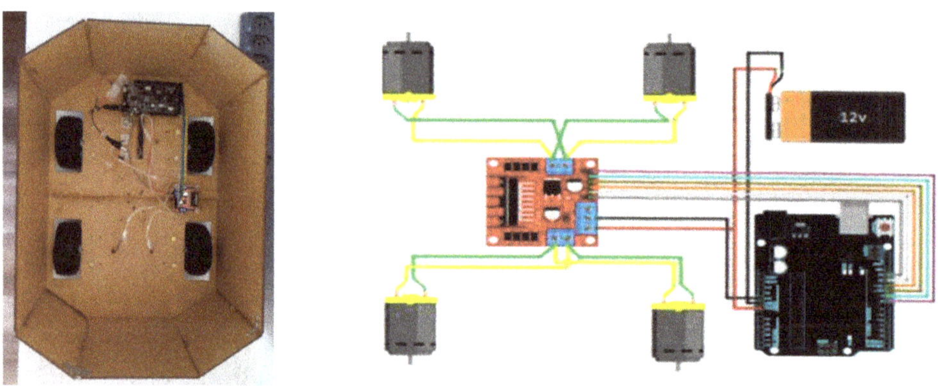

(A) (B)

Figure 8. Analysis of the motors in the fifth prototype: (**A**): distribution of internal components; (**B**): electrical diagram.

To carry out the movement test, a test track (Figure 9) was created in a commercial poultry shed during the empty period. This track was designed with 3.0 m of length and 60 cm of width. Furthermore, an additional load of 2.5 kg was added to simulate an approximate value for the weight of the environmental sensor equipment and the final robotic structure. For each type of motor, three uniform rectilinear movements were performed on the test track. Several authors propose different ways of evaluating robot movement, but in common they propose analysis on a specified (single) trajectory [31,32]. These authors do not define the number of movements necessary in this trajectory for evaluation. In the present study, it was opted to run the robot on a specified trajectory and with three rectilinear movements due to the imperfections of the chicken litter in the robot's displacement.

Figure 9. Locomotion track and motor testing.

What was evaluated in this test was the robot's ability to move in two aspects: (1) start and (2) constant movement until reaching the end of the track (for both cases, with and without additional load). For this purpose, a single motion control algorithm (same rotation value) applied to the different motors was created; the flowchart of this algorithm is presented in Figure 10. The result of the test allowed us to understand both the travel time between the beginning and end of the track and the performance of the motorization.

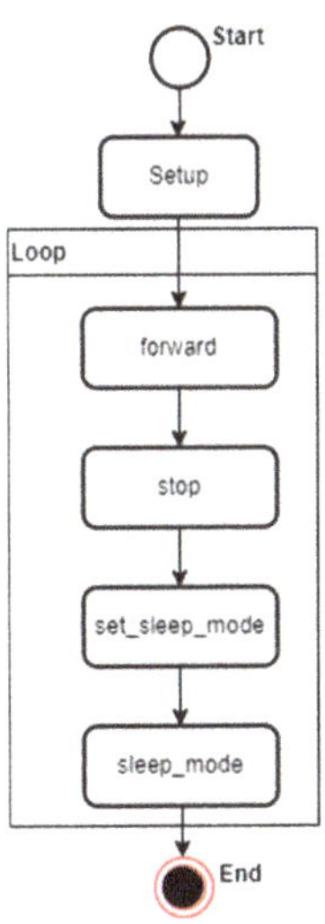

Figure 10. Robot movement algorithm.

The results of the robot's moving time (in minutes and seconds) in the four motorizations are presented in Table 4 (movement without additional load and movement with additional load).

Table 4. Travel time results with and without additional load.

Motors	With Additional Load						Without Additional Load					
	Essay			Averages			Essay			Averages		
	1 (s)	2 (s)	3 (s)	TT (s)	TTSD	TS m/s	1 (s)	2 (s)	3 (s)	TT (s)	TTSD	TS m/s
DC 80 rpm	25	26	29	26.67	1.73	0.11	-	-	-	-	-	-
N20 300 rpm	18	21	19	19.33	1.29	0.16	-	225	252	238.5	13.5	0.02
25 ga 370 12 rpm	100	96	99	98.33	1.73	0.03	200	204	202	202	1.63	0.01
25 ga 370 26 rpm	37	38	37	37.33	0.47	0.08	41	38	40	39.67	1.29	0.08

TT: travel time; TTSD: standard deviation; TS: travel speed.

The movement performed without additional load presented satisfactory results for all four motors as they were all able to move the robot structure from the starting point (0 m) to the end point (3 m). The N20 motor (300 rpm) showed the shortest time (0 min 19 s) to move with the highest speed (0.16 m/s), while the 25 ga 370 motor (12 rpm) showed the longest time (1 min 38 s) with a speed of 0.03 m/s, as expected due to its RPM.

The movement performed with additional load presented satisfactory results for the 25 ga 370 motors of 12 rpm and 26 rpm and unsatisfactory results for the DC 80 rpm and N20 300 rpm motors. This is due to the fact that the 80 rpm DC motor was unable to start the robot, remaining stationary in the attempts made, and the 300 rpm N20 motor was unable to start in the first attempt, succeeding in the others, but taking a very long time to complete the movement. Furthermore, after carrying out the test trials, one of the four N20 motors showed an oil leak, indicating that it exerted greater effort than expected, which probably damaged its internal structure [33].

It was concluded that the 12 rpm 25 ga 370 and 26 rpm 25 ga 370 motors were the only ones that managed to present satisfactory results in the tests carried out, making them suitable for execution in the poultry shed, since the 80 rpm DC motors were unable to move the robot with additional load, and the 300 rpm N20 motors were unable to complete one of the three tests with additional load. However, it is necessary to carry out new experiments to determine how much its speed of execution interferes with coexistence with chickens. Therefore, the next work proposed was an analysis of the "escape distance" of the animals (ED) in their different production phases, to determine how much these two motors impact their movement during operation in the aviary in avoiding crowding, and how they compromise animal welfare. It should be considered that among the motors evaluated, the speed varied between 0.03 m/s and 0.08 m/s (without additional load) and between 0.01 m/s and 0.08 m/s (with additional load).

Finally, the fifth prototype proved to be sufficient to move with stability presenting enough internal space to accommodate its components and environmental sensor module. However, the way it was designed caused it to be higher than the feeder lines (height of the robot: 30 cm; minimum height of the feeder lines: 22 cm), making it difficult to pass under the lines, having to go around them. Furthermore, it kept its motors exposed at the bottom, which made them subject to waste from the bedding, compromising their functioning. With these field limitations in accordance with the management of the aviaries, a new prototype was started.

3.1.6. Sixth Prototype

The sixth and final prototype (Figure 11) was built with a technological evolution that involved replacing the microcontroller with a microcomputer. Previous prototypes were sufficient to determine the battery, motorization and wheels. However, the complexity of the environment demonstrated the need to use more robust technologies, such as computer vision, to determine, for example, the method of moving the robot around the poultry shed. To do so, the Wemos Wifi ESP8266 + AtMega328p microcontroller was exchanged for the NVIDIA® Jetson Nano™ Developer Kit, which was added to a distance detection system based on a light sensor 360° camera (Light Detection and Ranging, model RP LiDAR A1; camera: Samsung model Np350xbe Ba96-07238a).

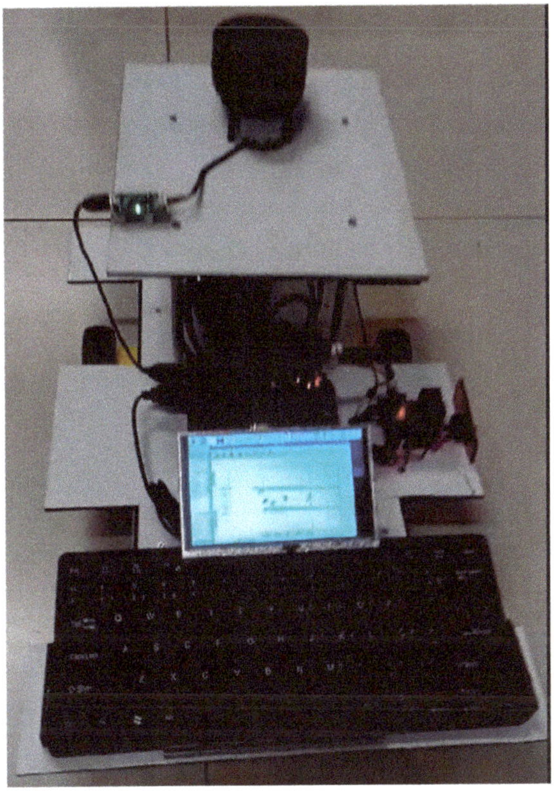

Figure 11. Sixth prototype: external view.

The use of LiDAR was sufficient to determine a 2D view of the distances of objects around the robot within a 360° radius. The LiDAR sensor emits pulses of light that strike objects and then detects the range of pulses that are reflected back. From the interval between the first and second pulse, it is possible to measure the time interval that the pulse takes to make this journey and thus calculate the distance in relation to the objects around it. Each LiDAR reading returns a vector with three elements (quality, angle, and distance). To determine the position of a reflected object within a map, the cosine value of position can be simply obtained.

The camera allowed for the visualization of the environment by recognizing objects using the OpenCV library in Python (version 3.13.0.). Specifically, it allowed for the real-time detection of chicken feeders with orange dishes (a limitation of this experiment, considering that there are dishes in other colors). However, this approach had a significant limitation: any other orange object would be identified as a feeder (Figure 12).

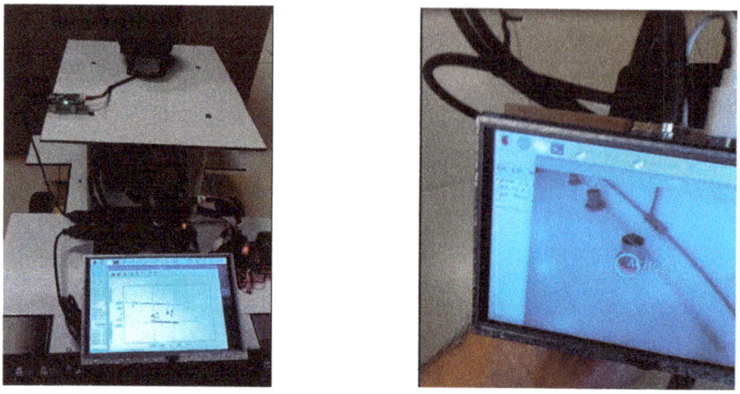

(**A**) (**B**)

Figure 12. Sixth prototype. (**A**): mapping of distances using LiDAR; (**B**): identification of feeders with computer vision.

3.2. Summary of Prototypes

Table 5 presents a summary of each prototype and the characteristics that each one provided to the project.

Table 5. Summary of each prototype and their characteristics.

Model	Prototype	Characteristics and Strengths
1	 https://youtu.be/aep_9DDyqn8 (accessed on 20 June 2025) *	The robot was built with a microcontroller and a 6 V DC motor driver, paired with 65 mm smooth wheels. However, it struggled with ground mobility due to insufficient motor power and poor traction. Additionally, the aluminum chassis posed electrical insulation issues, creating a risk of short circuits.
2		The second iteration used the same core technology but featured a plastic chassis with increased clearance. While this design resolved the electrical insulation issues, its compact size proved insufficient to hold all electronic components, leading to its prompt rejection.
3	 https://youtu.be/lKueFngYUpI (accessed on 20 June 2025) *	The third iteration maintained the same microcontroller technology but utilized an MDF chassis, offering excellent maneuverability due to its lightweight and easily modifiable structure. This design successfully validated the Ackermann steering geometry for directional control. However, during operation, debris accumulation in the wheel joints became problematic, leading to the prototype's discontinuation.

Table 5. *Cont.*

Model	Prototype	Characteristics and Strengths
4	 https://youtu.be/SFegoSuGxck (accessed on 20 June 2025) *	This iteration retained the same core technologies but was specifically designed to navigate beneath feeder and drinker lines. However, the design encountered two critical issues: animals began climbing onto the robot, and the lack of speed control caused distress among the birds during movement. These operational limitations led to the prototype's discontinuation.
5	 https://youtu.be/ylwnrjXJc1A (accessed on 20 June 2025) *	The adoption of rubber wheels and a higher-power motor enabled the construction of a larger robotic structure, effectively preventing animals from climbing onto it. However, this modification inadvertently heightened livestock distress, necessitating behavioral studies to analyze movement patterns and flight distances. These investigations aimed to determine the optimal motor RPM that would ensure both operational efficiency and animal welfare.
6	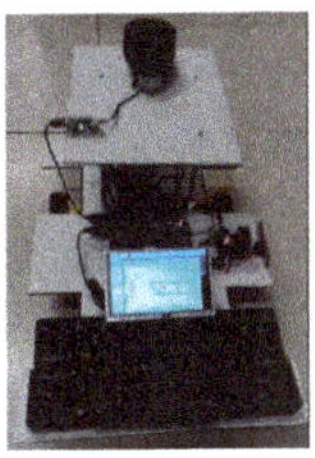 https://youtube.com/shorts/B59BfF8ZzNM (accessed on 20 June 2025) *	While retaining the same motor system, the microcontroller was upgraded to a microcomputer, which enabled the integration of higher-precision sensors, including computer vision and LiDAR, to monitor animal proximity and dynamically adjust the robot's movement speed. The enhanced system achieved stable operation while maintaining animal welfare by preventing distress responses.

* The video available at the link below shows the operation of the developed prototype, demonstrating its practical application and the results obtained through computer vision.

3.3. Modeling and Physical Implementation of *RobôFrango*

The acquired knowledge from the six prototypes developed was systematically integrated into the final *RobôFrango* design. This mobile robot was engineered to navigate poultry house bedding in the presence of livestock while transporting an environmental sensor module. The hardware architecture incorporated four critical capabilities: (1) autonomous locomotion, (2) precise maneuvering, (3) obstacle detection, and (4) computer vision for future route programming. The previous prototypes provided essential insights for these functional requirements. The environmental module of *RobôFrango* was designed for real-time monitoring of key poultry house parameters, including dry bulb temperature, relative humidity, gas concentrations (NH_3, CO, CO_2) and particulate matter levels.

The modeling of the robotic structure was carried out in the TinkerCad® software, and the approval was updated in AutoCAD 2020®, both from the company AutoDesk. The structure was based on the modeling proposed by [34], who divided the modeling of the environmental disinfection robot into four parts that allowed interdependence between the components—navigation compartment, disinfection unit, monitoring unit and intelligent control—with its modeling divided into a structure consisting of three parts (platform, body and trunk):

- **Platform:** supports the wheels, motors, batteries and microcomputer;
- **Body:** lateral structural parts supporting the robot and fixing the movement sensors (LiDAR) and camera;
- **Trunk:** allows the coupling of the sensory analysis device (environmental sensor module).

Aiming to distribute the processing workload and energy consumption, the robotic structure was designed as a scalable platform for electronic modules having its own energy source and specific microcomputer to carry out movements around the poultry shed. The environmental sensor module is a device separated from the robotic structure with its independent electrical and electronic circuit. The robotic structure serves as a vehicle that transports this device in the poultry shed and can be connected to it or be independent units. Dimensions and comparisons with existing robots are presented in Table 6.

Table 6. Dimensions of robots in poultry farming.

Robot	Length	Height	Width	Weight	Detailing
RobôFrango	58 cm	22 cm	36 cm	8.420 kg	Four motorized wheels, controlled by two L298N H bridge drivers
PoultryBot	110 cm	45 cm	55 cm	45.00 kg	Three motorized pneumatic wheels controlled by Roboteq AX3300 driver [30]
Lavicole	170 cm	450 cm	120 cm	1300 kg	Two pairs of conveyors supporting the weight of the engines that release water and detergent under high pressure, featuring a 22 hp Honda engine powered by gasoline [19]
Gasmetric	40 cm	40 cm	40 cm	U.I.	Extendable on steel cables that direct its autonomous movement throughout the aviary, recording environmental measurements in various vertical positions (descending from the ceiling to close to the floor) [17]
Spoutinc Nav	63 cm	18 cm	56 cm	12 kg	Four-wheel motorization controlled by driver controller [18]
ChickenBoy	60 cm	20 cm	25 cm	2.250 kg	Moved by rail suspended throughout the aviary [11]
OctopusRobot	140 cm	80 cm	112 cm	80 Kg	Four wheels powered by a 12 v battery (autonomy of up to 10 h) with autonomous or manual navigation [11]

U.I.: uninformed.

Figure 13 shows the external view of the sixth version of the robot. In (A), the front part of the structure is shown, which consists of a rubberized bumper to avoid contact with animals, with a LiDAR sensor above to detect feeder lines and a video camera below for computer vision (Samsung model Np350xbe Ba96-07238a). The environmental sensor module is attached to the upper base (luggage compartment). In (B), the front view is shown, where the LiDAR and the camera stand out. The bottom-rear view of the structure is shown with the presence of another rubberized bumper and the presence of the four wheels that are protected by a skirt that makes up the side of the robot.

The internal distribution of electronic components was achieved according to the spatial availability of the components and distribution in their electrical scheme and was guided by the robot's center of gravity. Figure 14 shows the electrical scheme of the motorization (prepared in the Fritzing software, version 0.9.3) and the modeling of the spatial distribution of these components in the 3D model.

(A) (B)

Figure 13. External view 3D modeling of the robotic structure: (**A**) front-side view and (**B**) bottom-rear view.

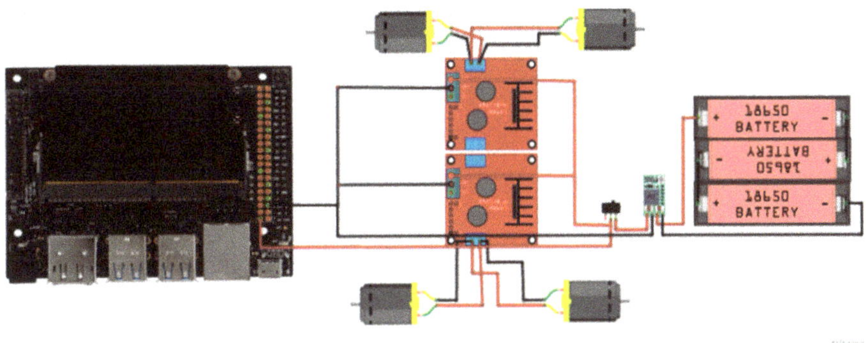

Figure 14. Internal view of the modeling of the robotic structure: electrical diagram of the motorization.

The 3D modeling made it possible to manufacture the robotic structure, which was made of MDF (Medium-Density Fiberboard) with a thickness of 3 mm and 6 mm (base for motors and wheels). The choice of this material was based on its malleability, low weight and low production cost. The dimensions of the 3D modeling were maintained on the laser cutting machine. The final product weighed 8420 kg (with batteries, motors, wheels, environmental module and electronic components included). Figure 15 shows the built prototype.

(A) (B) (C)

Figure 15. External view of the robotic physical structure: (**A**) front; (**B**) front without environmental sensor module compartment cover; (**C**) internal view.

The internal components were distributed according to the modeling presented in the electrical circuit and their physical implementation (Figure 16).

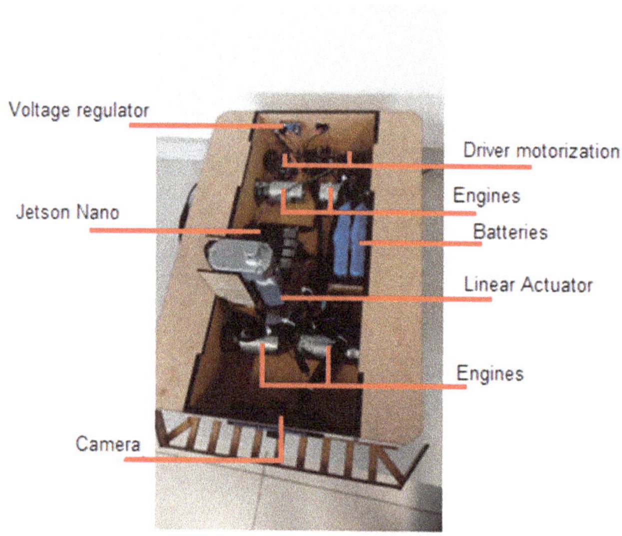

Figure 16. Distribution of electronic components.

Table 7 presents the electronic components (hardware) used in the construction of the robotic structure. Electronic components such as microcomputers, sensors and actuators were used. All software was processed by the NVIDIA® Jetson Nano™ Developer Kit microcomputer and programmed in the Python 3.11.2 programming language.

Table 7. Electronic components of the robotic structure motorization.

Component (Quantity)	Description
Microcomputer (1)	NVIDIA® Jetson Nano™ Developer Kit
Driver handling motors (2)	Bridge H L298N
Wheel motorization (4)	12 v motor of 26 rpm
Battery (4)	Li-Ion *Kedanone* 6 × 18,650 rechargeable 50,000 mAh
Wheels (4)	127 mm × 65 mm
Linear Actuator (1)	12 v 70 mm 1200 N
Humidity sensor (1)	GC-58
Computer vision camera (1)	Samsung Np350xbe Ba96-07238a
LiDAR Sensor (1)	A1M8, range 0.13 to 8 m, 360 degrees
Voltage regulator (1)	LM2596 input: 3.5–40 v; output: 1.5–35 v; 2A nominal, 3A maximum

3.4. Coding and Field Experiment

The operating system used was Ubuntu version 20.04.6 LTS (Focal Fossa) adapted for the Advanced RISC Machine (ARM) platform. Coding was implemented in the Python programming language (version 2.7.18). This language was chosen for the following reasons: (a) it is compatible with the NVIDIA® Jetson Nano™ Developer Kit microcomputer; (b) it has compatibility with a set of hardware manipulation libraries used in this project; and (c) it is compatible with an Integrated Development Environment (IDE) for professional use (Visual Studio Code version 1.88.1). The code construction method followed evolutionary prototyping followed by versioning. Coding was divided into four stages, as follows:

- **Motorization:** A program in Python language was coded to manipulate the four 12 v motors and their execution speeds using two HL298N bridges (one controlled the two wheels on the left side of the robot and the other the two wheels on the right side). Four functions that guided the movement of the robotic structure were built: forward, backward, left, right and stop movement. The control of the two HL298N bridges was performed through the RPi.GPIO library in which the basic motor control settings (forward or reverse) were defined using modularized functions in Python, as described in Table 8. Details of the code can be obtained at https://github.com/glauberbalthazar83/RoboFrango/blob/main/RoboFrango.py (accessed on 8 July 2025).
- **Distance detection:** The program was enhanced with the function of detecting distances from objects in their movement, for which the LiDAR sensor was used. Vroegindeweij et al. (2018) used a particle filter with a single LiDAR-type distance measurement sensor that identified the presence of obstacles in the trajectory; therefore, the trajectory was maintained, but obstacles were avoided. Distance analyses carried out in the literature review demonstrated that a distance of up to 20 cm was sufficient to avoid collisions with obstacles [30]. There are related studies that deal with different forms of navigation and obstacle avoidance techniques [35]. Ref. [36] presents a method using the LiDAR sensor that extracts spatial information from the point cloud laser using segmentation and grouping methods. The results of these studies suggest a distance of 0.20 m to 0.30 m up to 100 m. These studies corroborate [30]. Therefore, LiDAR was used in this project with a minimum distance of 20 cm from objects in front of it.
- **Coloration (robot painting):** Colors, particularly light ones, due to their varying wavelengths, can affect the behavior, welfare and performance of birds [37–39]. Generally, chickens prefer the color white [40] as these animals have greater sensitivity in spectra from 350 nm to 700 nm, and the color white has wavelengths between 400 nm and 700 nm. According to [40], the color white represents a high stimulus for animals when perceiving the environment, as it stimulates the birds' entire field of vision [41,42]. Thus, the robot was painted white.
- **Controlled movement:** The movement of the robotic structure was guided by the rows of animal feeder lines and by the measurements obtained by the LiDAR sensor. It was determined that the movement of the robotic structure would be centered on maximum and minimum limits of distance from the feeder line: maximum between 45 and 55 cm and minimum between 25 and 35 cm. These distances were chosen by observing attempts to move the robotic structure between the feeder lines in order to maintain a safe distance from the drinker line (parallel to the feeder line) and the feeder plates (schematic drawing in Figure 17). According to the specifications presented by [43], who also determined left and right navigability safety measures based on readings from ultrasonic sensors, it was possible to establish the travel path as an ideal distance range for the feeder line between 30 and 50 cm $\pm$ 5 cm. The LiDAR sensor captured the actual distances from the robot to the feeder line, from which it was possible to make a decision about which movement the robot should perform. This decision could be made manually by a human being via remote control (Remote Desktop Connection software—Microsoft) or autonomously, with the robot deciding on its own which movement is most appropriate according to the calculations of the median distances to the feeder line, as shown in Table 9.
- **Battery analysis:** The following current consumption tests were performed: when turning on the robot (motor tests), starting consumption was 0.38 mAh; during loading of the operating system (initialization and loading of the Python program with the motor

manipulation drives), it was 0.67 mA; after finishing loading of the operating system and robot drives and in standby, consumption was 0.82 mA; during movement, the current consumption was measured when performing the robot's displacement movements: forward: 3.66 mA; to the left: 2.48 mA; to the right: 2.36 mA; and backward: 3.68 mA. Considering the battery capacity used of 50,000 mAh and a continuous rectilinear movement of the robot forward, we divided its capacity by the forward consumption (50,000 mAh/3.66 mA), obtaining an autonomy of 13,661 h. However, in practice, it only obtained ~6 h of continuous execution. The factors that may have led to this difference may have been the poultry house bedding having many undulations, forcing the motors to work at maximum power to move the robot; the battery may have a lower actual capacity than specified; despite the battery being new, it may have presented accelerated self-discharge problems (excessive loss of charge when used in extreme situations); and lithium batteries lose capacity in very hot and humid environments, such as poultry houses (which may cause more pronounced degradation).

Table 8. Python code for integration with HL298N bridge (hardware).

Function	Modularized Functions of Control Movement

Movement stop	`def stop():`
	`print("stop")`
	`GPIO.output(in1,GPIO.LOW)`
	`GPIO.output(in2,GPIO.LOW)`
	`GPIO.output(in3,GPIO.LOW)`
	`GPIO.output(in4,GPIO.LOW)`

Movement forward	`def toForward():`
	`print("forward")`
	`GPIO.output(in1,GPIO.LOW)`
	`GPIO.output(in2,GPIO.HIGH)`
	`GPIO.output(in3,GPIO.HIGH)`
	`GPIO.output(in4,GPIO.LOW)`

Movement backward	`def toBackward():`
	`print("backward")`
	`GPIO.output(in1,GPIO.HIGH)`
	`GPIO.output(in2,GPIO.LOW)`
	`GPIO.output(in3,GPIO.LOW)`
	`GPIO.output(in4,GPIO.HIGH)`

Movement to left	`def toLeft():`
	`print("left")`
	`GPIO.output(in1,GPIO.LOW)`
	`GPIO.output(in2,GPIO.HIGH)`
	`GPIO.output(in3,GPIO.LOW)`
	`GPIO.output(in4,GPIO.LOW)`

Movement to right	`def toRight():`
	`print("right")`
	`GPIO.output(in1,GPIO.LOW)`
	`GPIO.output(in2,GPIO.LOW)`
	`GPIO.output(in3,GPIO.HIGH)`
	`GPIO.output(in4,GPIO.LOW)`

Three tests were carried out to move the robotic structure from the starting point to the end of the aviary (traversing the three partitions) and the robot was executed by remote controlling. A video is available at https://youtube.com/shorts/f-g73piGydw (accessed on 20 June 2025). The following average times were obtained (average times for each partition present strong fluctuations due to the different stages of the litter in each part of the shed) (hh:mm:ss):

- First partition (45 m): 00:25:32 (average speed: 0.030 m/s);
- Second partition (45 m): 00:23:52 (average speed: 0.032 m/s);

- ○ Third partition (45 m): 00:24:02 (average speed: 0.031 m/s);
- ○ Total time (135 m): 01:13:26 (final average speed: 0.031 m/s).

Table 9. Python code for autonomous decision to move.

```
for t in scan:
  angle = int(t[1]) #iterations read angle sensor LiDAR
  if(angle<360):
    x[angle] = int(t[2]) * math.cos(math.radians(angle))
    y[angle] = int(t[2]) * math.sin(math.radians(angle))
  if(t[1]>=335 or t[1]<=25):  #50° from an angle forward
    values_forward.append(t[2])
  if(t[1]>=155 and t[1]<=205):  #50° from an angle to the back
    values_backward.append(t[2])
  if(t[1]>=206 and t[1]<=324):  #118° from an angle to the right
    values_right.append(t[2])
  if(t[1]>=26 and t[1]<=154):  #118° from an angle to the left
    values_left.append(t[2])
# statistics (median of set angle) determine decision to direction of robot
  if len(values_forward)>0:
    med_forward = statistics.median(values_forward)
if(med_forward>=20):
    toForward()
  if len(values_backward)>0:
    med_backward = statistics.median(values_backward)
if(med_backward >=20):
    toBackward()
  if len(values_right)>0:
    med_right = statistics.median(values_right)
if(med_right >=45 and med_right <=55):
    toRight()
  if len(values_left)>0:
    med_left = statistics.median(values_left)
if(med_left >=25 and med_left <=35):
    toLeft()
```

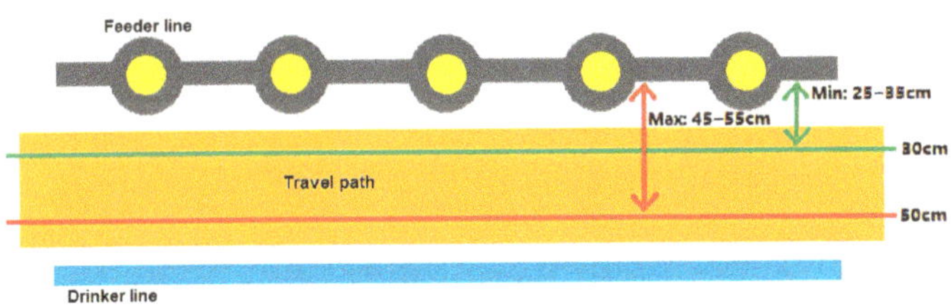

(A)

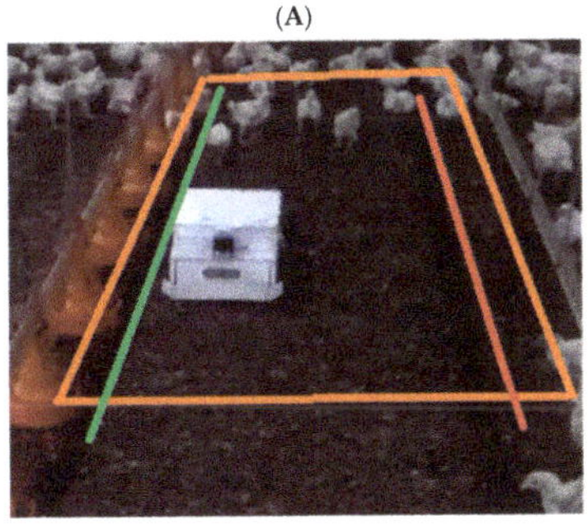

(B)

Figure 17. Modeling the navigation distances of the robotic structure guided by the feeder line: (**A**) description of distances and (**B**) field vision (green: minimum of 25–35 cm; red: maximum of 45–55 cm). Note: The orange trapezoid outlines the region of interest (ROI) for spatial navigation modeling.

4. Conclusions

The hypothesis of this study was proven considering that it was possible to build a mobile robotic structure using fast electronic prototyping equipment, allowing it to be executed in a commercial broiler aviary. Therefore, it was possible to determine the following components for the robotic structure: 130 mm wheels, with a 42 mm tire width rubberized with claws; a Li-Ion battery with 12 v voltage and 50,000 mAh charging capacity; a 12 v 25 ga 370 DC motor (12–25 kgf/cm) controlled by an L298N H bridge; an RP1 LiDAR sensor; and a Samsung camera, model Np350xbe Ba96–07238a. The execution algorithm was developed and implemented on the robot, enabling it to operate within the poultry farm. The robot successfully traversed the entire length of the aviary, and its components proved to be robust without suffering damage.

The incorporation of mobile robotics into broiler production can enhance efficiency, productivity, and animal welfare, contributing to the sustainability and competitiveness of the poultry sector. The benefits for broiler production include continuous monitoring of house conditions (e.g., temperature, humidity, lighting), enabling rapid adjustments to optimize animal performance and welfare; automated assistance with repetitive tasks, such as collecting dead animals; early detection of health or behavioral issues in chickens, allowing for swift and effective intervention; collection of production data, including weight gain, feed conversion, and mortality rate, for analysis and strategic decision-making; and integration with farm management systems, providing real-time information to optimize overall production efficiency.

Limitations of the study: This stage of development of *RobôFrango* was limited to analyzing the mechanical and electronic components for the construction of the robot's chassis and electrical circuit. Additional studies were carried out to evaluate the robot's locomotion condition considering the animals' welfare, through tests of escape distance speed. The results are demonstrated by da Rocha Balthazar [44].

Author Contributions: G.d.R.B.: data collection, writing—original draft, conceptualization, data curation, methodology, writing—original draft, validation; R.M.F.S.: writing—original draft, validation; I.J.O.d.S.: data collection, writing—original draft, conceptualization, data curation, methodology, writing—original draft, validation. All authors have read and agreed to the published version of the manuscript.

Funding: This research was funded by the São Paulo State Research Support Foundation (FAPESP), grant number 2022/07442-8, and the Luiz de Queiroz Agricultural Studies Foundation (FEALQ).

Institutional Review Board Statement: The animal study protocol was approved by the Ethics Committee on the Use of Animals of the University of São Paulo (protocol code: 7364090322; date of approval: 16 June 2021).

Informed Consent Statement: Not applicable.

Data Availability Statement: The datasets generated and/or analyzed during the current study are available upon request to the corresponding author.

Conflicts of Interest: The authors declare no conflicts of interest.

References

1. Kopler, I.; Marchaim, U.; Tikász, I.E.; Opaliński, S.; Kokin, E.; Mallinger, K.; Neubauer, T.; Gunnarsson, S.; Soerensen, C.; Phillips, C.J.C.; et al. Farmers' Perspectives of the Benefits and Risks in Precision Livestock Farming in the EU Pig and Poultry Sectors. *Animals* **2023**, *13*, 2868. [CrossRef]
2. Tzanidakis, C.; Simitzis, P.; Panagakis, P. Precision Livestock Farming (PLF) Systems: Improving Sustainability and Efficiency of Animal Production. In *International Series in Operations Research & Management Science*; Springer: Cham, Switzerland, 2023; pp. 285–337, ISBN 9783031166198. [CrossRef]

3. Franzo, G.; Legnardi, M.; Faustini, G.; Tucciarone, C.M.; Cecchinato, M. When Everything Becomes Bigger: Big Data for Big Poultry Production. *Animals* **2023**, *13*, 1804. [CrossRef]
4. Depuru, B.K.; Putsala, S.; Mishra, P. Automating Poultry Farm Management with Artificial Intelligence: Real-Time Detection and Tracking of Broiler Chickens for Enhanced and Efficient Health Monitoring. *Trop. Anim. Health Prod.* **2024**, *56*, 75. [CrossRef] [PubMed]
5. Bao, J.; Xie, Q. Artificial Intelligence in Animal Farming: A Systematic Literature Review. *J. Clean. Prod.* **2022**, *331*, 129956. [CrossRef]
6. Wu, D.; Cui, D.; Zhou, M.; Ying, Y. Information Perception in Modern Poultry Farming: A Review. *Comput. Electron. Agric.* **2022**, *199*, 107131. [CrossRef]
7. Singh, M.; Kumar, R.; Tandon, D.; Sood, P.; Sharma, M. Artificial Intelligence and IoT Based Monitoring of Poultry Health: A Review. In Proceedings of the 2020 IEEE International Conference on Communication, Networks and Satellite (Comnetsat), Batam, Indonesia, 17–18 December 2020; IEEE: New York, NY, USA. [CrossRef]
8. Yuan, J.; Ji, W.; Feng, Q. *Robots and Autonomous Machines for Agriculture Production*; MDPI: Basel, Switzerland, 2023. Available online: https://mdpi-res.com/bookfiles/book/7738/Robots_and_Autonomous_Machines_for_Agriculture_Production.pdf?v=1710859530 (accessed on 25 March 2024).
9. Uzedhe, G.O.; Akinloye, B.O.; Febaide, I.C. Development of an Animal Farm Robotic Feeding System. *Trop. J. Sci. Technol.* **2023**, *4*, 14–22. [CrossRef]
10. Parajuli, P.; Huang, Y.; Tabler, T.; Purswell, J.L.; DuBien, J.L.; Zhao, Y. Comparative Evaluation of Poultry-Human and Poultry-Robot Avoidance Distances. *Trans. ASABE* **2020**, *63*, 477–484. [CrossRef]
11. Ren, G.; Lin, T.; Ying, Y.; Chowdhary, G.; Ting, K.C. Agricultural Robotics Research Applicable to Poultry Production: A Review. *Comput. Electron. Agric.* **2020**, *169*, 105216. [CrossRef]
12. Park, M.; Britton, D.; Daley, W.; McMurray, G.; Navaei, M.; Samoylov, A. Artificial intelligence, sensors, robots, and trans-portation systems drive an innovative future for poultry broiler and breeder management. *Anim. Front.* **2022**, *12*, 147–162. [CrossRef]
13. Priva, M.; Pavithraa, K.; Devi, P.; Sureshkumar, V. Smart Poultry Farm Incorporating GSM and IoT. *Int. Res. J. Eng. Technol.* **2021**, *8*, 3186–3190. Available online: https://www.irjet.net/archives/V8/i3/IRJET-V8I3613.pdf (accessed on 22 April 2024).
14. Chang, C.-L.; Xie, B.-X.; Wang, C.-H. Visual Guidance and Egg Collection Scheme for a Smart Poultry Robot for Free-Range Farms. *Sensors* **2020**, *20*, 6624. [CrossRef]
15. Montalcini, C.M.; Voelkl, B.; Gómez, Y.; Gantner, M.; Toscano, M.J. Evaluation of an Active LF Tracking System and Data Processing Methods for Livestock Precision Farming in the Poultry Sector. *Sensors* **2022**, *22*, 659. [CrossRef] [PubMed]
16. Olejnik, K.; Popiela, E.; Opaliński, S. Emerging Precision Management Methods in Poultry Sector. *Agriculture* **2022**, *12*, 718. [CrossRef]
17. Gasmetric. Gasmetric Lanza un Nuevo Producto de Medición del Ambiente en Granjas. Available online: https://seleccionesavicolas.com/files/digital/revista-de-avicultura-seleccionesavicolas/2016-05-selecciones-avicolas-revista/html5/index.html?page=047 (accessed on 23 January 2023).
18. Spoutinic Nav. Invented by Breeders, Developed for Their Needs, Rewarded by Their Results. 2016. Available online: https://tibot.fr/ (accessed on 23 January 2023).
19. Lavicole Robot. Self-Propelled and Radio-Controlled Cleaning Machine Designed for Poultry Houses, Chicken Coops. Available online: https://www.rabaud.com/en/products/poultry-cleaning-machine/poultry-houses-cleaning-machine-lavicole.html (accessed on 23 January 2023).
20. Özentürk, U.; Chen, Z.; Jamone, L.; Versace, E. Robotics for Poultry Farming: Challenges and Opportunities. *Comput. Electron. Agric.* **2024**, *226*, 109411. [CrossRef]
21. Edan, Y.; Adamides, G.; Oberti, R. Agriculture Automation. In *Springer Handbook of Automation*; Springer: Cham, Switzerland, 2023; pp. 1055–1078, ISBN 9783030967284. [CrossRef]
22. Sassi, B.N.; Averós, X.; Estevez, I. Technology and Poultry Welfare. *Animals* **2016**, *6*, 62. [CrossRef]
23. Lei, T.; Li, G.; Luo, C.; Zhang, L.; Liu, L.; Gates, R.S. An Informative Planning-Based Multi-Layer Robot Navigation System as Applied in a Poultry Barn. *Intell. Robot.* **2022**, *2*, 313–332. [CrossRef]
24. Haixia, Q.; Jinzhuo, J.; Chaohai, W. Path Planning for Agricultural Robots in Wild Livestock Farm Environments. *Int. J. Agric. Biol. Eng.* **2024**, *17*, 207–216. [CrossRef]
25. Quan, Q.; Palaoag, T.D.; Sun, H. Research and Design of Intelligent Inspection Robot for Large-Scale Chicken Farms. In Proceedings of the 2024 5th International Conference on Machine Learning and Human-Computer Interaction (MLHMI), Kawasaki, Japan, 14–16 March 2024; IEEE: New York, NY, USA. [CrossRef]
26. Yang, D.; Cui, D.; Ying, Y. Development and Trends of Chicken Farming Robots in Chicken Farming Tasks: A Review. *Comput. Electron. Agric.* **2024**, *221*, 108916. [CrossRef]
27. Zhang, Y.; Sun, W.; Yang, J.; Wu, W.; Miao, H.; Zhang, S. An Approach for Autonomous Feeding Robot Path Planning in Poultry Smart Farm. *Animals* **2022**, *12*, 3089. [CrossRef]

28. Shutherland, J.; Shutherland, J.J.; Lua, N. *Scrum: A Arte de Fazer o Dobro do Trabalho na Metade do Tempo*, 1st ed.; Sextante: Rio de Janeiro, Brazil, 2019.
29. Łebkowski, A. Temperature, Overcharge and Short-Circuit Studies of Batteries Used in Electric Vehicles. *Prz. Elektrotech.* **2017**, *1*, 69–75. [CrossRef]
30. Vroegindeweij, B.A.; Blaauw, S.K.; IJsselmuiden, J.M.M.; van Henten, E.J. Evaluation of the Performance of PoultryBot, an Autonomous Mobile Robotic Platform for Poultry Houses. *Biosyst. Eng.* **2018**, *174*, 295–315. [CrossRef]
31. Garganta, G.R.B. *Protótipos de Robots Móveis com Diferentes Configurações de Locomoção*, 1st ed.; Faculdade de Engenharia da Universidade do Porto: Porto, Portugal, 2023.
32. Fu, C.; Jing, B.A.I.; Huiyu, W.; Xiaoguang, X.I.E. Reducing the Energy Consumption of a Palletizing Robot through RobotStudio. In Proceedings of the 2019 Chinese Control and Decision Conference (CCDC), Nanchang, China, 3–5 June 2019; IEEE: New York, NY, USA.
33. Eletronics Pololu Robotics & Micro Metal Gearmotor. Pololu. 2018. Available online: www.pololu.com/product/1098 (accessed on 18 March 2021).
34. Feng, Q.C.; Wang, X. Design of Disinfection Robot for Livestock Breeding. *Procedia Comput. Sci.* **2020**, *166*, 310–314. [CrossRef]
35. Pandey, A. Mobile Robot Navigation and Obstacle Avoidance Techniques: A Review. *Int. Robot. Autom. J.* **2017**, *2*, 1–12. [CrossRef]
36. Ghorpade, D.; Thakare, A.D.; Doiphode, S. Obstacle Detection and Avoidance Algorithm for Autonomous Mobile Robot Using 2D LiDAR. In Proceedings of the 2017 International Conference on Computing, Communication, Control and Automation (ICCUBEA), Pune, India, 17–18 August 2017; IEEE: New York, NY, USA, 2017.
37. Genc, M.; Ozenturk, U. The Effect of Worker Clothing Color on Stress in Laying Hens. *Arch. Anim. Breed.* **2024**, *67*, 145–151. [CrossRef]
38. Nelson, J.R.; Bray, J.L.; Delabbio, J.; Archer, G.S. Light Emitting Diode (LED) Color and Broiler Growth: Effect of Supplementing Blue/Green LED to White LED Light on Broiler Growth, Stress, and Welfare. *Poult. Sci.* **2020**, *99*, 3519–3524. [CrossRef]
39. Li, G.; Li, B.; Zhao, Y.; Shi, Z.; Liu, Y.; Zheng, W. Layer Pullet Preferences for Light Colors of Light-Emitting Diodes. *Animal* **2019**, *13*, 1245–1251. [CrossRef]
40. Siriani, A.L.R. *Referência de Galinhas Para Diferentes Comprimentos de Onda de luz Usando Deep Learning*, 1st ed.; Universidade Estadual Paulista Júlio de Mesquita Filho: Tupã, Brazil, 2022.
41. Riber, A.B. Effects of Color of Light on Preferences, Performance, and Welfare in Broilers. *Poult. Sci.* **2015**, *94*, 1767–1775. [CrossRef]
42. Prescott, N.B.; Wathes, C.M. Spectral Sensitivity of the Domestic Fowl (Gallus g. Domesticus). *Br. Poult. Sci.* **1999**, *40*, 332–339. [CrossRef]
43. Johnson, J.; Godwin, D.J. Indoor Navigation of Mobile Robot Using Fuzzy Logic Controller. In Proceedings of the 2015 3rd International Conference on Signal Processing, Communication and Networking (ICSCN), Chennai, India, 26–28 March 2015; IEEE: New York, NY, USA, 2015.
44. da Rocha Balthazar, G.; Silveira, R.M.F.; da Silva, I.J.O. How Do Escape Distance Behavior of Broiler Chickens Change in Response to a Mobile Robot Moving at Two Different Speeds? *Animals* **2024**, *14*, 1014. [CrossRef]

AgriEngineering

MDPI

Article

Effect of Green Roofs on the Thermal Environment of Prototype Broiler Houses

Maria Angela de Souza [1], Fernanda Campos de Sousa [1,*], Alex Lopes da Silva [2], Thauane Cordeiro Soares [2], Charles Paranhos Oliveira [1], Ricardo Brauer Vigoderis [3], Fernando da Costa Baêta [1] and Ilda de Fátima Ferreira Tinôco [1]

[1] Department of Agricultural Engineering, Federal University of Viçosa, Viçosa 36570-900, MG, Brazil; maria.a.souza@ufv.br (M.A.d.S.); charles.paranhos@ufv.br (C.P.O.); baeta@ufv.br (F.d.C.B.); iftinoco@ufv.br (I.d.F.F.T.)
[2] Department of Animal Science, Federal University of Viçosa, Viçosa 36570-900, MG, Brazil; alex.lopes@ufv.br (A.L.d.S.); thauane.soares@ufv.br (T.C.S.)
[3] Department of Agricultural Engineering, Federal University of Agreste of Pernambuco, Garanhuns 55292-270, PE, Brazil; ricardo.vigoderis@ufape.edu.br
* Correspondence: fernanda.sousa@ufv.br; Tel.: +55-31998952772

Abstract: The management of thermal environments in animal production facilities presents significant challenges, requiring continuous adjustments to meet animals' physiological needs. This study evaluated the effects of green roofs on the thermal environment and comfort indices in small-scale poultry house prototypes, comparing facilities with and without green roof installations. The research tested various roof types (ceramic, fiber cement, and metal) combined with emerald grass (*Zoysia japonica*) green roof systems. Parameters measured included air temperature, relative humidity, internal roof surface temperature, Temperature and Humidity Index (THI), Black Globe Humidity Index (BGHI), Human Comfort Index (HCI), and Thermal Radiation Load (TRL) under both open and closed conditions. Results showed that green roofs reduced indoor air temperature by up to 2.4 °C in open prototypes and 10.6 °C in closed prototypes during peak heat periods. In combinations using green roofs with fiber cement tiles, internal roof surface temperature decreased by 24.0 °C in open prototypes and 27.0 °C in closed configurations. The implementation of green roofs resulted in THI reductions of 2.3 and 8.1 units in open and closed prototypes, respectively, BGHI decreases of 2.8 and 11.3 units, and TRL reductions of 21.0 W/m^2 and 74.0 W/m^2. HCI measurements confirmed improved thermal comfort conditions with green roof installations in both settings. This study concludes that green roofs effectively enhance the thermal environment by reducing bioclimatic indices during hot periods while maintaining stable conditions during cooler weather, thereby improving overall thermal comfort in animal facilities.

Keywords: animal ambience; animal welfare; rural constructions; thermal stress

1. Introduction

Brazilian poultry farming stands out in the animal production sector through the continuous development and implementation of advanced technologies that enhance productive efficiency while reducing operational costs [1]. With an accumulated growth exceeding 1.087% in chicken meat production, Brazil has established itself as the second-largest producer and primary global exporter, solidifying its leadership position in the international market [2]. However, despite the robust expansion of the poultry sector, significant knowledge gaps persist regarding thermal conditioning in tropical poultry facilities, particularly concerning energy consumption, heating and cooling systems efficiency, animal welfare, and air quality [3]. Consequently, while addressing the increasing demand for animal protein, the fundamental challenge of maintaining optimal thermal comfort for animals throughout the production cycle remains paramount. Thermal comfort constitutes a critical factor for both welfare and productive performance in intensive farming systems [4,5].

In response to accelerated climate change, regions experience elevated temperatures, increased precipitation variability, and intensified frequency of extreme weather events. The convergence of climate change impacts and escalating livestock production demand needs to increase production sustainably while minimizing environmental impacts [6–8]. Considering climate's substantial influence on animal production, implementing innovative construction alternatives and ambiance strategies becomes essential to ensure thermal and energy efficiency in facilities [9,10]. These strategies encompass strategic shading implementation, selection of high thermal capacity materials for heat regulation, and adoption of advanced heating mitigation technologies. Such comprehensive measures optimize internal thermal conditions, enhancing facility sustainability while minimizing environmental impact and energy consumption, simultaneously ensuring animal welfare. Within this framework, agricultural facilities' roofing emerges as a crucial element, playing an instrumental role in controlling solar radiation and buildings' thermal load [4,11]. To establish an optimal thermal environment, a comprehensive understanding of roofing materials' thermal properties becomes essential, facilitating enhanced heat exchange dynamics in buildings [12,13].

In this context, green roof technology represents an innovative construction approach that has gained significant recognition in scientific research and the construction sector due to its comprehensive social, environmental, and economic benefits [14]. Green roof systems serve as an effective strategy for optimizing buildings' thermal comfort, offering substantial advantages in both hot and cold climates. In hot climates, these systems effectively reduce internal thermal load, while in cold climates, they facilitate heat retention, thereby maintaining optimal thermal comfort [15–17]. The moisture retention capacity of green roof systems plays a fundamental role in reducing surface temperature and enhancing cooling efficiency through evapotranspiration, particularly during summer periods. Cooling efficiency is predominantly influenced by substrate thickness and vegetation cover density. The plant canopy functions as a natural radiation shield, significantly reducing solar radiation transmission to the roof surface, consequently lowering the surface temperature and diminishing building cooling requirements. The canopy structure further modulates heat exchange processes, as foliar surfaces reflect, transmit, and absorb solar radiation, thereby regulating the system's thermal dynamics [18]. Beyond its thermal insulation properties, the green roof system orchestrates solar radiation balance through convection, evaporation, and heat conduction mechanisms, reflecting 20% to 30% of solar radiation, using up to 60% for photosynthetic processes, and facilitating evaporative cooling. The system's thermal efficiency correlates with substrate depth variations while additionally contributing significantly to rainwater management optimization [19–22].

Evidence suggests that green roof systems possess considerable potential for mitigating animal production impacts within the context of global climate change while simultaneously reducing the effects of elevated air temperatures on animal production systems. However, existing literature predominantly focuses on green roof applications in urban buildings [10,16,18,19,21,23–30]. This urban-centric focus constrains understanding of green roof potential in alternative contexts, particularly animal production environments [31,32]. Implementing green roof systems in rural settings, specifically animal production facilities, offers significant advantages, including enhanced thermal comfort and improved sustainability metrics. Beyond their established positive effects on building thermal conditioning, green roof systems contribute substantially to energy conservation through reduced heating and cooling system demands. Through investigating this technology in a relatively unexplored context, this study provides empirical evidence demonstrating green roofs' impacts on energy efficiency and animal welfare in rural environments. Thermal comfort assessment in agricultural facilities relies on specialized indexes that quantify comfort conditions for both animals and human operators. Among these comprehensive measures, the Temperature and Humidity Index (THI) serves as a primary indicator, evaluating the synergistic effects of temperature and humidity [33,34]. Additionally, the Black Globe Humidity Index (BGHI) incorporates environmental thermal radiation measurements [35], while the Human Comfort Index (HCI) evaluates human well-being through temperature–humidity interactions [36]. To maintain optimal occupational conditions, thermal comfort parameters must adhere to standards established by the Ministry of Labor and Employment [37]. Furthermore, the Radiant Thermal Load (RTL) provides a quantitative measurement of direct radiation impacts on biological systems [38]. These thermal comfort indexes enable comprehensive monitoring of green roofs' effects on thermal comfort conditions for both animals and human operators within production systems.

Furthermore, the utilization of reduced-scale models enables precise prediction of thermal conditions in full-scale installations, accurately simulating environmental thermal behavior. This scaled approach proves particularly valuable for complex investigations, such as those focusing on thermal comfort in animal production, as it facilitates experimental control, optimizes research efficiency, and enhances understanding of underlying physical principles [13,39–41]. Thus, this research investigates green roof effects on the thermal environment of reduced-scale poultry house prototypes, specifically characterizing the thermal environment and thermal comfort indices in poultry house prototypes with and without green roof systems.

2. Materials and Methods

2.1. Experimental Location

This investigation was conducted at the Department of Agricultural Engineering's (DEA) Rural Constructions and Environment Area, Federal University of Viçosa (UFV), located in Viçosa/MG, Brazil. The experiment was carried out from 24 February to 11 March 2022 at the Research Center on Ambience and Engineering of Agro-Industrial Systems (AMBIAGRO), positioned at latitude 20° 45′ south and longitude 45° 52′ west, with an elevation of 670 m. According to the Köppen classification, the local climate is characterized as Cwa, featuring a subtropical pattern with dry, cold winters and hot, humid summers. The experimental phase was conducted during the summer season. Six scale-reduced commercial poultry house prototypes were constructed following a 1:5 scale similarity concept [42]. The prototypes featured identical dimensions: 2.5 m in width and 3.2 m in length, with a ceiling height of 0.6 m and an eave of 0.5 m. These structures were distributed in two parallel rows and positioned on level, grass-covered terrain with an east–west orientation. The eaves of the prototypes were closed with wood.

The poultry houses' prototypes were evaluated under field conditions, without external environmental control, due to the open nature of the experimental area. The experiment was conducted in two distinct phases: in the first, the prototypes were left with an open construction design, while in the second, blue polyethylene side curtains were installed, allowing partial enclosure of the structure. Each experimental condition was maintained for a period of seven days, during which data were systematically collected.

2.2. Vegetation Coverture

Emerald grass (*Zoysia japonica*) represents one of the predominant species used in Brazil for landscaping applications, sports facilities, and ornamental gardens, owing to its exceptional adaptability to tropical climate conditions [43,44]. This species demonstrates superior resistance to foot traffic and minimal maintenance requirements, making it extensively used for ground coverage in transportation infrastructure, public spaces, and residential areas, where its aesthetic appeal and durability are highly valued [45]. These characteristics enhance its suitability for green roof applications [31,46], where it contributes to thermal insulation, water retention, and resource efficiency. The increasing implementation of ornamental grasses in urban and residential environments has established *Zoysia japonica* as the primary selection for landscaping initiatives throughout the country [43].

2.3. Irrigation Management of Vegetation Coverture

Water requirements of the grass coverage were determined based on crop evapotranspiration (ETc) and calculated using the Penman–Monteith method (Equation (1)), following Allen et al. [47]. Reference evapotranspiration (ET0) values were derived from climatic variables (air temperature, relative humidity, atmospheric pressure, wind speed, and radiation) obtained from INMET [48].

$$\mathrm{ETc} = \mathrm{ET0} \times \mathrm{Kc} \quad (1)$$

where ETc represents crop evapotranspiration, ET0 represents reference evapotranspiration, and Kc represents the crop coefficient.

A Kc value of 1 was implemented for *Zoysia japonica* grass, as established by Madeira et al. [49], given its utilization as a reference species in the evapotranspiration determination model. Reference evapotranspiration (ET0) calculations incorporated climate variables from the preceding irrigation day. This estimation enabled the calculation of crop evapotranspiration for the total grass coverage area, thereby determining daily water requirement, which represented the volume of water applied per day accounting for daily precipitation in the available water depth (Table 1). Manual irrigation was conducted throughout the experimental period, initiating one day before each experimental data collection phase and occurring at three daily intervals, 08:00 A.M., 1:00 P.M., and 6:00 P.M., in accordance with daily water requirements.

Table 1. Daily estimation of water demand for *Zoysia japonica* grass during the summer in Viçosa-MG, based on crop evapotranspiration (ETc) and precipitation.

Day	ETc (mm)	Water Demand ($L{\cdot}m^{-2}{\cdot}dia^{-1}$)	Precipitation (mm)
1	4.51	40.59	0.20
2	4.35	39.15	0.00
3	4.23	38.07	0.00
4	4.39	39.51	0.00
5	4.55	40.95	8.40
6	4.46	40.14	0.40
7	4.22	37.98	0.00
8	4.81	43.29	0.00
9	4.53	40.77	0.00
10	4.45	40.05	0.00
11	4.55	40.95	0.00
12	4.50	40.50	0.00
13	4.29	38.61	0.20
14	4.55	40.95	0.20
15	3.29	29.61	1.40
16	4.50	40.50	0.00

Source: Authors, based on INMET data [48].

2.4. *Types of Tested Roofing Materials*

The experimental treatments comprised three different roofing material combinations, evaluated both with and without green roof systems. The roofing materials tested included Roman ceramic tile (CT), corrugated fiber cement tile (FCT), and trapezoidal metallic tile (MT). Three of the six prototypes incorporated each roofing material type in conjunction with a green roof system using *Zoysia japonica* grass. The investigation encompassed two experimental phases evaluating distinct building typologies: open and closed prototypes. The closed prototype configuration featured blue polyethylene curtains installed along the longitudinal sides of the structures (Figure 1).

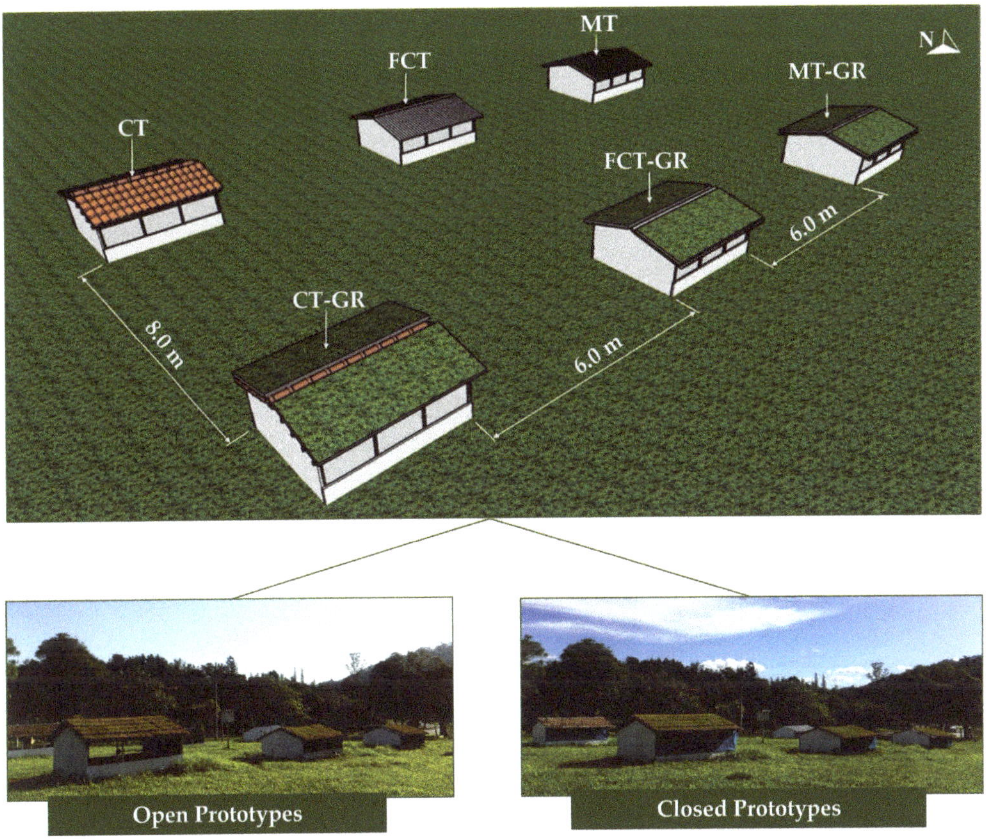

Figure 1. Representation of the positioning of prototypes in the experimental area, with different treatments: prototype with ceramic tile (CT), prototype with fiber cement tile (FCT), prototype with metal tile (MT), prototype with ceramic tile and green roof (CT-GR), prototype with fiber cement tile and green roof (FCT-GR), and prototype with metal tile and green roof (MT-GR). Emphasizing the different constructive typologies, with open prototypes and closed prototypes. Source: Authors.

Before the experimental period, ceramic and fiber cement tiles underwent thorough cleaning procedures. Subsequently, these materials received varnish waterproofing treatment to prevent moisture infiltration. The green roof installation process included the use of a black shade cloth, which served as a support for the substrate, facilitating the uniform distribution of the grass over the roof surface. Each prototype had a coverage area of 9 m^2, with a thickness of 8 cm/m^2 and a grass density of 26 kg/m^2. In addition, the fiber cement tiles used in this study had moderately darkened surface characteristics due to previous exposure, which may have increased their heat absorption capacity compared to new materials.

2.5. Characterization of the Thermal Environment of the Prototypes

The internal thermal environment of each prototype underwent systematic monitoring through continuous temperature and relative humidity measurements. Monitoring used Onset® dataloggers (model HOBO UX100-03; Onset Computer Corporation, Bourne, MA, USA), equipped with temperature measurement capabilities ranging from −20.0 °C to +70.0 °C (accuracy: ±0.21 °C; resolution: ±0.024 °C) and relative humidity measurement capabilities spanning 15.0% to 95.0% (accuracy: ±3.5%; resolution: 0.07%). Two sensors, positioned at each prototype's geometric center, facilitated continuous internal environment monitoring. Data acquisition occurred at minute intervals throughout the 24 h cycle during the entire experimental period. One sensor recorded dry-bulb temperature (air temperature) and relative humidity, enabling dew point temperature calculation, while a second sensor, positioned within a black globe, measured black globe temperature. Additionally, air velocity measurements were obtained at each prototype's geometric center using a TAFR-200 hot-wire anemometer (measurement range: 0.1 to 25.0 m/s; resolution: 0.01 m/s; accuracy: 5%; Wöhler Technik GmbH, Bad Wünnenberg, Germany). For treatment analysis evaluating various covering materials with and without green roof systems, internal surface temperature measurements of prototype roofs were obtained using a Fluke® infrared thermometer (model 62 MAX; Fluke Corporation, Everett, WA, USA). This instrument featured a measurement range of –30.0 °C to +500.0 °C (accuracy: ±1.5 °C; resolution: 0.1 °C). Surface temperature determination occurred from an internal geometric reference point, measured at 0.5 m distance, with emissivity adjustments corresponding to specific covering material (Table 2). The measurements were conducted at 3 h intervals throughout the 24 h cycle during the experimental period.

Table 2. Emissivity values of the surface temperature sensor of different roofing materials.

Coverture Material	Emissivity (£)
Ceramic tile	0.85–0.95
Fiber Cement Roof Tile	0.92
Metal Roof Tile	0.25
Green Roof (Grass)	0.90

Source: Adapted from Baêta and Souza [11].

The characterization of the external thermal environment of the prototypes in the experimental area was performed by determining the air temperature and relative humidity values throughout the entire experimental period using data from the National Institute of Meteorology [48]. Descriptive statistical analyses were performed to calculate the mean, minimum, and maximum values of air temperature and relative humidity in the external environment.

2.6. Characterization of Thermal Comfort Indexes

The effects of treatments, encompassing various roofing materials with and without green roof systems, on the prototype's thermal comfort, were evaluated using established bioclimatological indexes. These metrics, extensively validated and widely implemented in scientific investigations [31,36,50–56], enable precise analysis of thermal comfort conditions for both animals and workers in animal production facilities. Their demonstrated effectiveness in characterizing poultry environments underscores their relevance in assessing covering material impacts and optimizing thermal conditions in poultry facilities. The THI was calculated using the model developed by Thom [33] and refined by Mader et al. [34], incorporating air temperature and relative humidity parameters (Equation (2)).

$$THI = 0.8 \times T_d + (RH/100) \times (T_d - 14.4) + 46.4 \quad (2)$$

where T_d represents the dry bulb temperature (°C), and RH represents the relative humidity of the air (%).

The (BGHI) was determined using Equation (3), established by Buffington et al. [35], using black globe temperature and dew point temperature measurements.

$$BGHI = T_{bg} + 0.36\ T_{dp} + 41.5 \quad (3)$$

where T_{bg} represents the black globe temperature (°C), and T_{dp} represents the dew point temperature (°C).

The HCI developed by Rosenberg [36] was determined and calculated using Equation (4), incorporating air temperature, relative humidity, vapor pressure (Equation (5)), and saturated air vapor pressure (Equation (6)).

$$HCI = T_a + \frac{5}{9}(e_a - 10) \quad (4)$$

$$e_a = \frac{(e_s \times RH)}{100} \quad (5)$$

$$e_s = 0.611 \times 10^{\left\lfloor \frac{7.5 \times T_a}{237.3 + T_a} \right\rfloor} \quad (6)$$

where T_a represents the air temperature (°C), RH represents the relative humidity of the air (%), e_a represents the vapor pressure (kPa), and e_s represents the saturated air vapor pressure (kPa).

The RTL established by Kelly and Bond [38] was determined based on the radiant thermal exchanges of the roof and calculated using Equation (7) through the Stefan–Boltzmann constant and the mean radiant temperature (MRT) values, as presented in Equation (8).

$$RTL = \sigma MRT^4 \quad (7)$$

where σ represents the Stefan-Boltzmann constant, $W \cdot (m^2 \cdot K^4)^{-1}$, and MRT represents the mean radiant temperature, (K).

$$MRT = 100\left(2.51\sqrt{V}\left(T_{bg} - T_d\right) + \left(\frac{T_{bg}}{100}\right)^4\right)^{\frac{1}{4}} \quad (8)$$

where V represents the air velocity ($m \cdot s^{-1}$), T_{bg} represents the black globe temperature (K), and T_d represents the dry bulb temperature (K).

To assess the thermal comfort of animals and humans, the calculated values of air temperature, relative humidity, THI, BGHI, and HCI were compared with the comfort ranges described in the literature, as presented in Table 3.

Table 3. Thermal comfort ranges for broilers (air temperature, relative humidity, THI, and BGHI) and workers (HCI).

Comfort Range	Week of Life					Authors
	1	2	3	4	5	
Air temperature (°C)	32.3–35.0	26.4–32.0	21.6–29.0	21.6–26.0	20.0–23.0	Silva et al. [51] and Oliveira et al. [53]
Relative humidity (%)	56.90	64.40	68.50	68.50	69.20	Oliveira et al. [53]
THI	72.4–80.0	68.4–76.0	64.5–72.0	60.5–68.0	56.6–64.0	Silva et al. [51]
BGHI	81.30	74.90	69.80	69.80	68.70	Oliveira et al. [53]
HCI			20.0–29.0 *			Santos [57], Souza et al. [58] and Santos et al. [59]

* Values for humans. Source: Adapted by authors.

2.7. Statistical Analyses

The entire dataset underwent statistical analysis using the Generalized Least Squares (GLS) procedure implemented in R statistical software version 4.1.3 [59]. The analytical framework incorporated the evaluation of multiple factors: roofing tile type, presence or absence of green roof systems, data collection timing, and their respective interactions. The temporal component of data collection was integrated into the model as a repeated measure, with an evaluation of multiple covariance structures, including compound symmetry, heterogeneous compound symmetry, first-order autoregressive, and heterogeneous first-order autoregressive matrices. The selection of the optimal covariance matrix structure was determined based on the minimization of the Akaike Information Criterion (AIC). The heterogeneous first-order autoregressive matrix demonstrated a superior fit and was subsequently implemented across all analyses. Mean separation procedures, when warranted by significant effects, were conducted using Tukey's multiple comparison test at a 0.05 probability threshold. In instances where significant interactions were detected, these were subjected to detailed decomposition analysis.

3. Results

3.1. Open Prototyping Environments

3.1.1. External Environment

The external environment in the open prototype experimental area was characterized using temperature and relative humidity data from INMET [48] over the seven-day experimental period, allowing the characterization of external thermal variations (Figure 2). The data revealed a well-defined daily cycle: air temperatures are lowest at dawn (16.0 °C to 19.0 °C), gradually increasing throughout the morning and reaching peak values between 26.0 °C and 30.0 °C from 12 P.M. to 3 P.M. Conversely, relative humidity exhibited an opposite pattern, with the high values at dawn (95% to 98%) and decreasing to 40% to 60% during the hottest afternoon hours (3 P.M.). This inverse relationship between temperature and relative humidity indicates that the external environment becomes drier as air temperature rises, particularly during the afternoon hours.

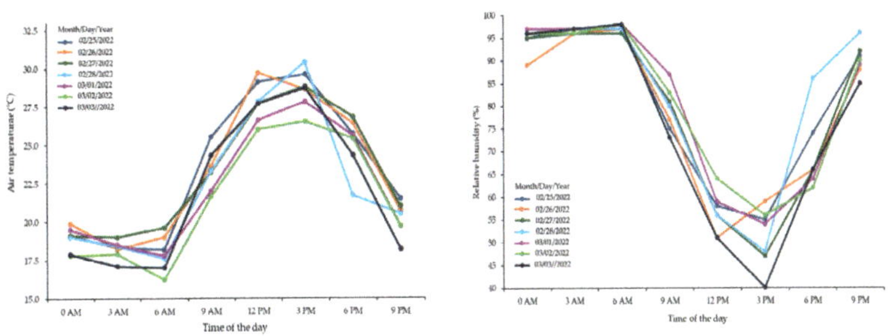

Figure 2. Characterization of the external thermal environment with temperature and relative humidity data during the experimental period with open prototypes in Viçosa-MG. Source: Adapted from INMET [48].

3.1.2. Thermal Environment

The open prototype experiments revealed significant differences (p-value < 0.0001) across all analyzed variables, including air temperature, internal surface temperature, and relative humidity, for the three types of roof tiles tested, both with and without green roofs, throughout the observation period (Figure 3). Statistical analysis of air temperature data showed a significant (p-value < 0.0001) simple interaction effect of time and a double interaction effect between green roof presence and time (Figure 3A).

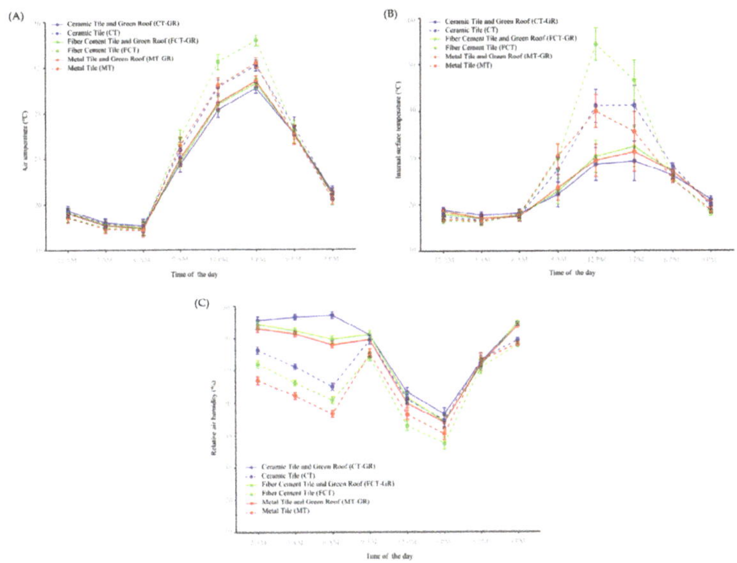

Figure 3. Behavior of air temperature (**A**), internal surface temperature (**B**), and relative humidity (**C**) in open prototypes, showing different types of roof tiles, with and without the presence of a green roof, at different times. Source: Authors.

Evaluation of internal surface temperature in open prototypes (Figure 3B) revealed a significant simple interaction effect (p-value < 0.0001) for both time and green roof presence, along with a double interaction between these factors. Similarly, relative humidity data indicated a significant interaction effect (p-value < 0.0001) for time, roof tile type, and green roof presence (Figure 3C), with additional double interactions observed between green roof and time and between roof tile type and time. These findings demonstrate the influence of roof tile type and green roof presence on thermal regulation and humidity levels under open prototype conditions.

Air temperature analysis in open prototypes, both with and without green roofs, revealed statistically significant differences at 9:00 A.M., 12:00 P.M., and 3:00 P.M., with higher air temperature recorded in prototypes without green roofs (Figure 3A). This finding highlights the substantial impact of green roofs on the thermal environment, particularly during peak external temperature periods when increased energy consumption is typically required to maintain optimal thermal conditions. Figure 3A demonstrates that prototypes with green roofs exhibited lower air temperatures compared to those without, showing an average reduction of 2.4 °C at both 12:00 P.M. and 3:00 P.M. during the hottest parts of the day.

Internal surface temperature data on roofing materials in open prototypes (Figure 3B) showed statistically significant differences (p-value < 0.0001) only at 3:00 P.M. and 9:00 P.M. The highest internal surface temperature was recorded in the prototype with fiber cement tiles without a green roof. The presence of green roofs resulted in substantial temperature reductions: 24.0 °C for fiber cement tiles, 12.6 °C for ceramic tiles, and 10.5 °C for metallic tiles.

Relative humidity analysis open prototypes (Figure 3C) revealed statistically significant differences (p-value < 0.0001) at most time points, with the exception of 9:00 A.M., 6:00 P.M., and 9:00 P.M. The highest relative humidity values were observed in prototypes with ceramic tiles during the cooler periods (9:00 P.M. to 6:00 A.M.), while the lowest values were recorded in prototypes with fiber cement tiles during the warmer hours (9:00 A.M. to 6:00 P.M.). Notably, prototypes with green roofs consistently maintained higher relative humidity levels, remaining above 70.0% between 9:00 P.M. and 6:00 A.M., corresponding with lower air temperatures.

3.1.3. Thermal Comfort

Open prototype experiments demonstrated significant variations (p-value < 0.0001) in thermal comfort indexes among the three types of roof tiles tested, with and without green roofs, throughout the observation period (Figure 4).

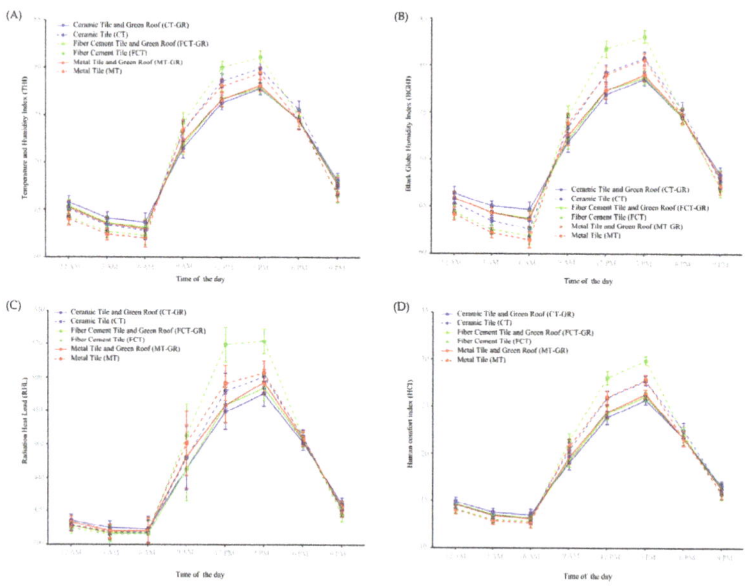

Figure 4. The behavior of the THI (**A**), BGHI (**B**), RTL (**C**), and HCI (**D**) in open prototypes demonstrating the effects of different types of roofing tiles, both with and without the presence of a green roof, at various time points. Source: Authors.

Statistical analysis of THI data in open prototypes (Figure 4A) revealed significant effects (p-value < 0.0001): a simple interaction with times of day and a double interaction between green roof presence and time of day. Similarly, BGHI data analysis (Figure 4B) showed significant simple interaction effects with time of day and double interaction effects between green roof presence and time of day (p-value < 0.0001). RTL data analysis (Figure 4C) identified significant simple interaction effects with time of day (p-value < 0.0001) and double interaction effects between green roof presence and time of day. HCI analysis in open prototypes (Figure 4D) revealed significant simple interaction effects with time of day and green roof presence, along with significant double interaction effects (p-value < 0.0001). These results demonstrate how temporal variation and green roof presence influence thermal comfort indexes in prototypes. THI analysis in open prototypes revealed statistically significant differences at 9:00 A.M., 12:00 P.M., and 3:00 P.M., with higher values in prototypes without green roofs, particularly those with fiber cement tiles (Figure 4A). At 3:00 P.M., green roof presence contributed to a THI reduction of 2.3 across all tile types. During nighttime hours (9:00 P.M. to 6:00 A.M.), prototypes with green roofs across all had higher average THI values than those without.

BGHI analysis in open prototypes showed that green roof presence did not produce statistically significant differences at 6:00 P.M. and 9:00 P.M. However, during peak temperature periods (12:00 P.M. and 3:00 P.M.), prototypes without green roofs exhibited elevated BGHI values, particularly those with fiber cement tiles (Figure 4B).

RTL data analysis in open prototypes revealed statistically significant differences (p-value < 0.0001) at 12:00 P.M. and 3:00 P.M. Prototypes showed that green roofs showed higher RTL averages during peak heat periods (9:00 A.M. to 3:00 P.M.), particularly those with fiber cement roofing (Figure 4C). During nighttime hours (6:00 P.M. to 6:00 A.M.), ceramic tiles with green roofs provided superior thermal insulation, maintaining higher RTL values compared to other treatments.

HCI analysis in open prototypes showed significant differences (p-value < 0.0001) at 9:00 A.M., 12:00 P.M., and 3:00 P.M. regarding green roof presence. For tile type, significant differences (p-value < 0.0001) were observed at 3:00 P.M. Prototypes without green roofs exhibited higher average HCI values during peak heat periods, with fiber cement tiles showing the highest HCI (Figure 4D). During nighttime (6:00 P.M. to 6:00 A.M.), ceramic tiles with green roofs demonstrated enhanced thermal insulation, maintaining higher HCI averages than other treatments.

3.2. *Closed Prototyping Environments*

3.2.1. External Environment

The external environment in the closed prototype experimental area was characterized using temperature and relative humidity data from INMET [48] over the seven-day experimental period, enabling the description of external thermal variations (Figure 5). Air temperature data showed minimum values at dawn, ranging from 17.5 °C to 21.2 °C, while peak values occurred in the afternoon between 26.1 °C and 30.0 °C. Relative humidity reached its highest levels in the early morning hours, ranging from 92% to 98%, and decreased throughout the day, reaching minimum values of 47% to 66% during peak temperature periods (12 P.M. and 3 P.M.). These data demonstrate an inverse relationship between temperature and relative humidity, where temperature increases correspond to decreases in relative humidity levels.

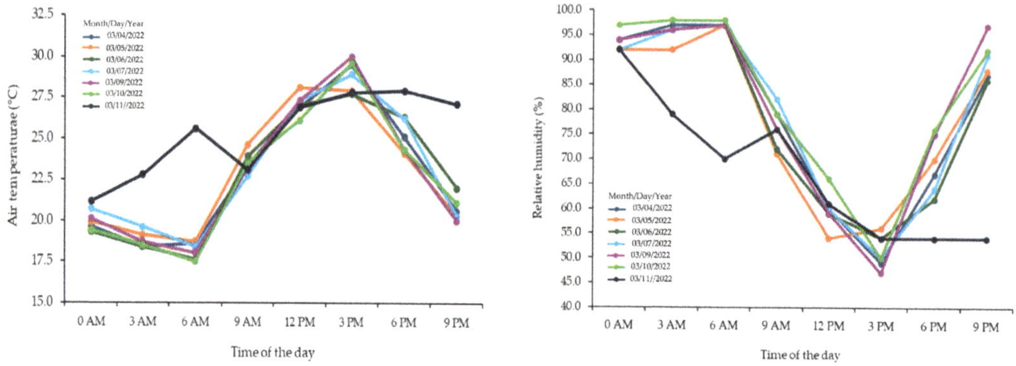

Figure 5. Characterization of the external thermal environment with temperature and relative humidity data during the experimental period with closed prototypes in Viçosa-MG. Source: Adapted from INMET [48].

3.2.2. Thermal Environment

The closed prototypes showed significant differences (p-value < 0.0001) across all analyzed thermal variables, including air temperature, internal surface temperature, and relative humidity. These differences were observed among the three types of roof tiles, both with and without green roofs, throughout the observation period (Figure 6).

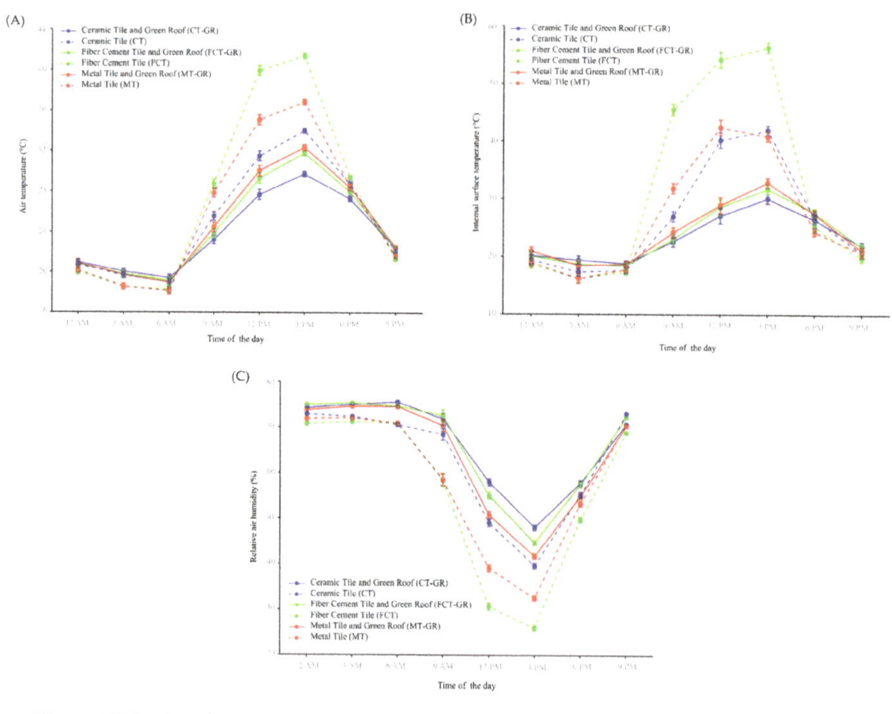

Figure 6. Behavior of air temperature (**A**), internal surface temperature (**B**), and relative humidity (**C**) in closed prototypes, showing different types of roof tiles, with and without the presence of a green roof, at different times. Source: Authors.

Statistical analysis of air temperature data from the closed prototype experiment (Figure 6A) revealed significant simple, double, and triple interactions (p-value < 0.0001).

Similarly, internal surface temperature analysis (Figure 6B) showed significant effects (p-value < 0.0001) across all interaction types. The relative humidity data analysis (Figure 6C) also demonstrated significant interactions (p-value < 0.0001) among all variables. Consequently, each interaction was examined individually to thoroughly investigate significant effects across different time points.

In closed prototypes (Figure 6A), significant interactions (p-value < 0.0001) were observed among all variables regarding internal air temperature values. The presence of green roofs resulted in substantial reductions in mean air temperature, particularly during peak heat periods. At 12:00 P.M., temperature reductions were 10.6 °C for fiber cement tiles, 5.1 °C for metallic tiles, and 3.8 °C for ceramic tiles. While no statistically significant temperature differences were found at 12:00 A.M. and 6:00 A.M., green roofs provided thermal damping during cooler periods (9:00 P.M. to 6:00 A.M.), with notable effects at 3:00 A.M., particularly in prototypes with metal tiles.

Surface temperature analysis in closed prototypes (Figure 6B) revealed significant interactions (p-value < 0.0001) at 9:00 A.M. and 3:00 P.M. Green roof presence significantly reduced surface temperatures at 3:00 P.M.: 24.6 °C for fiber cement tiles, 8.0 °C for metallic tiles, and 11.9 °C for ceramic tiles. While no significant differences were observed at 9:00 P.M., 12:00 A.M., and 6:00 A.M., green roofs exhibited a damping effect during cooler hours, with a statistically significant difference at 3:00 A.M.

Relative humidity analysis in closed prototypes (Figure 6C) showed significant interactions (p-value < 0.0001) among all variables, with statistically significant differences at all time points. Green roof presence led to substantial increases in mean relative humidity, particularly at 3:00 P.M.: 18.7% in prototypes with fiber cement tiles, 9.2% with metal tiles, and 8.5% with ceramic tiles. These increases can be attributed to the prototypes' closed lateral sides, which limited natural ventilation and consequently affected thermal exchanges.

3.2.3. Thermal Comfort

The closed prototype experiments revealed significant differences (p-value < 0.0001) in thermal comfort indexes across all three roof tile types, both with and without green roofs, throughout the observation period (Figure 7).

Statistical analysis of THI data (Figure 7A) showed significant simple, double, and triple interaction effects (p-value < 0.0001) across nearly all analyzed variables. Similarly, BGHI data analysis (Figure 7B) revealed significant effects (p-value < 0.0001) from simple interactions with time, double interactions between tile type and time, green roof and time, as well as triple interactions among time, tile type, and green roof presence. RTL data analysis in closed prototypes (Figure 7C) demonstrated significant effects (p-value < 0.0001) in simple, double, and triple interactions across all tested variables. HCI data analysis (Figure 7D) also demonstrated significant simple, double, and triple interaction effects (p-value < 0.0001) among all variables. Each interaction was analyzed separately to evaluate effects across different times of day.

In closed prototypes (Figure 7A), significant interactions were observed across almost all analyzed variables. At 3:00 P.M., green roof presence led to substantial THI reductions: 7.0 for fiber cement tiles, 3.8 for ceramic tiles, and 3.5 for metal tiles. Prototypes without green roofs exhibited the highest average THI values at 12:00 P.M. and 3:00 P.M., particularly those with fiber cement and metal tiles, attributed to the thermal properties of these materials. Conversely, prototypes with green roofs maintained lower THI values, especially those with ceramic tiles.

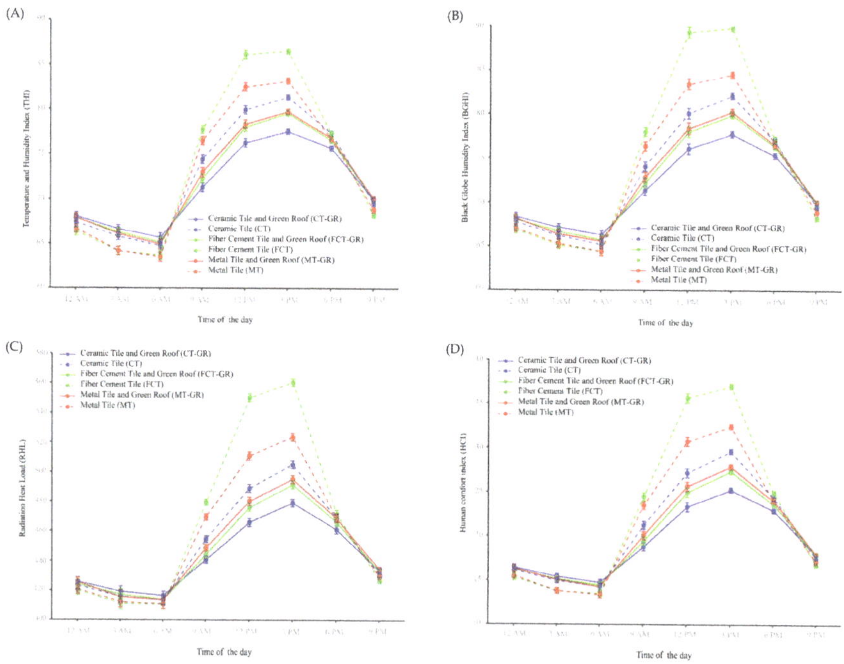

Figure 7. Behavior of the THI (**A**), BGHI (**B**), RTL (**C**), and HCI (**D**) in closed prototypes, demonstrating the effects of different types of roofing tiles, both with and without the presence of a green roof, at various time points. Source: Authors.

Closed prototype analysis (Figure 7B) revealed significant interactions (p-value < 0.0001) among almost all variables. Green roof presence resulted in substantial reductions in average BGHI values, particularly at 3:00 P.M., with the most pronounced reductions in prototypes with fiber cement tiles. During cooler times (6:00 P.M. to 06:00 A.M.), green roofs provided BGHI damping effects, though values remained higher compared to prototypes without green roofs. RTL analysis in closed prototypes (Figure 7C) showed significant interaction effects (p-value < 0.0001) among all variables. Green roof implementation reduced average RTL values during peak daytime temperatures, providing enhanced protection against incident solar radiation. HCI analysis in closed prototypes (Figure 7D) demonstrated interactions (p-value < 0.0001) among all variables. Prototypes with green roofs maintained lower HCI values and superior thermal conditions between 09:00 A.M. and 6:00 P.M. compared to those without green roofs.

4. Discussion

Given the characteristic climatic conditions at the experimental site and the varying experimental conditions of the prototypes (Figures 2 and 5), there is a clear need to adjust the thermal environment of the facilities to meet animal requirements, ensuring their thermal comfort and well-being [60]. Maintaining appropriate thermal comfort conditions is fundamental for efficient broiler production, requiring continuous adjustments in environmental variables [61,62]. Thermal management must adapt to both bird development and regional climatic variations. During the initial weeks, birds undergo environment acclimatization, while in later growth stages, increased body mass elevates the internal temperature of the shed, necessitating more stringent thermal control [62].

Furthermore, it is important to highlight that the ambient temperature is influenced not only by the roof but also by the walls of the installation. The characteristics of the walls, such as material and thickness, impact the retention and dissipation of heat, affecting the internal thermal environment. The side curtains of the prototypes, by functioning as a barrier to natural ventilation, allow the evaluation of treatments under controlled ventilation, providing a more precise analysis of the impact of thermal control on the internal environment.

When evaluating the broiler's life cycle in a real facility, it can be seen that the air temperature inside the prototypes (Figure 3A) remains at thermal comfort levels (Table 3) throughout the day, especially in the prototypes with the green roof. This configuration proved to be effective in mitigating heat gain, contributing to more stable indoor conditions suitable for animal welfare. On the other hand, at the hottest times of the day, the prototypes with fiber cement roofs had internal temperatures above the comfort limits, showing the unfavorable thermal impact of this material. Fiber cement roof tiles have intrinsic characteristics that limit their thermal performance, such as a thickness of 0.008 m, a specific heat of 0.84 kJ/kg.K, a density of 1900 kg/m^3, and a heat capacity of 12.77 kJ/(m^2.K) [63]. These factors result in lower thermal inertia, leading the material to absorb and transfer heat quickly, in contrast to ceramic and metal tiles, which have greater thermal stability. In addition, the aging of fiber cement intensifies this effect since the darkening of the material increases its solar radiation absorption coefficient from 0.45 in new tiles to 0.80 in aged tiles [4,64]. This increase in heat absorption increases thermal discomfort, especially in environments with high solar incidence. During the critical hours of the day, the air temperature data in the open prototypes (Figure 3A) confirmed that fiber cement had the highest values, surpassing ceramic and metal tiles. This trend is attributed to the lower thermal inertia of fiber cement, which contributes significantly to the thermal instability observed [65].

The results observed in this study align with those found by Barnabé et al. [66], who reported similar air temperatures in prototypes with fiber cement roofing during peak heat hours, ranging from 32.0 to 35.0 °C between 12:00 P.M. and 3:00 P.M. This is consistent with the findings in northeastern Brazil, where they recorded maximum air temperatures of approximately 31.8 °C in calf housing with fiber cement roofing. Likewise, Oliveira et al. [53] evaluated fiber cement roofing in reduced-scale poultry house models and observed air temperatures around 33.0 °C, further confirming the trend observed in the present study. However, when comparing different roofing materials, Oliveira et al. [53] found that ceramic tiles provided superior solar radiation interception, with temperatures ranging from 30.6 °C to 31.6 °C, compared to fiber cement and metal tiles without additional roof protection. This supports the notion that ceramic tiles are more effective at functioning as thermal barriers, contributing to lower internal temperatures during peak heat periods. The findings from Oliveira et al. [53] and the present study emphasize the role of ceramic tiles in improving thermal comfort in animal production facilities, offering better insulation and minimizing temperature fluctuations, which can enhance animal welfare and productivity. In contrast, the higher air temperatures observed in prototypes with fiber cement roofing during peak heat hours suggest that this material is less effective in providing thermal insulation. This discrepancy may be attributed to the material's heat absorption properties, which are influenced by factors such as color, surface texture, and thermal mass. As a result, adjusting roofing materials or integrating additional protective measures, such as green roofing or reflective coatings, could be beneficial in improving the thermal performance of fiber–cement roofs and reducing internal temperatures.

Air temperature results in prototypes with lateral enclosure using ceramic tiles and green roofing (27.7 ± 0.5 °C), measured at 12:00 P.M. (Figure 5), were comparable to those reported by Carneiro et al. [31]. In their evaluation of green roofs in Pernambuco state using prototype poultry sheds with *Zoysia japonica* grass and peanuts (Arachis repens), they observed average internal air temperatures of approximately 27.7 °C for both plant species. Emphasizing the decisive role of selecting roofing materials that meet thermal comfort requirements is crucial, as they directly affect the internal microclimate and, consequently, the well-being and productivity of the animals housed. Green roofs, in particular, excel at stabilizing temperature fluctuations. By creating a more consistent indoor environment, they mitigate heat stress and provide optimal conditions for animal production, even under reduced ventilation. The comparison with the study by Carneiro et al. [31] reinforces the potential of green roofs as a sustainable and effective solution for improving thermal regulation in animal housing facilities.

In the present study [31], air temperature varied between 18.0 ± 0.3 °C and 30.5 ± 0.2 °C over the period of study in open prototypes with green roofing (Figure 3A), enabling an environment suitable for adult birds, which ranges between 18.0 and 28.0 °C in summer conditions with natural ventilation [11]. However, it is important to emphasize that the ideal thermal comfort conditions vary with the age of the birds. The selection of roofing materials must take thermal requirements into account in order to guarantee a favorable internal microclimate, directly impacting animal welfare and productivity. It should be noted that between the third and fourth week of production (Table 3), the comfortable air temperature range is 23.0 to 29.0 °C [67,68]. In this context, the green roof is an effective construction technique, promoting thermal comfort and internal temperature during these production phases.

The elevated air temperature inside the prototypes with side closure (Figure 6A) results in reduced air exchange between internal and external environments. This reduction allows the energy from incident solar radiation to increase internal air mass temperature, which remains trapped due to the lateral enclosures' limitation of air exchange. However, green can significantly reduce internal air temperature [69], as the vegetative cover minimizes temperature gains during periods of solar exposure.

Abreu et al. [67] established representative thermal comfort ranges for poultry throughout their reproductive cycle (Table 3): 29.0 to 32.0 °C in the second week and 20.0 to 23.0 °C in the fifth week. These ranges align with those observed in the present study, with the lowest values recorded in prototypes featuring green roofing and ceramic tiles. Air temperature analysis in prototypes revealed variation between 16 °C and 44 °C over the 24 h observation period (Figures 3A and 6A). The green roof prototypes maintained temperature within human comfort limits, according to La Roche et al. [16], who categorize thermal comfort in human dwellings into three levels: below 21.1 °C (cold), between 21.1 °C and 25.5 °C (comfortable), and above 25.5 °C (warm). The recorded temperature values indicate an environment with adequate thermal comfort for workers, conforming to Ministry of Labor and Employment standards [37].

Internal surface temperature analysis in the prototypes (Figures 3B and 6B) demonstrates that green roofing can significantly reduce thermal load in facilities, promoting a more suitable thermal environment for livestock. Pragati et al. [70] corroborate this finding, indicating that surface temperatures in buildings with green roofs are approximately 5.17% lower compared to conventional buildings without greening systems. Carneiro et al. [31] observed surface temperatures of approximately 36.7 °C on green roofs with *Zoysia japonica* grass, demonstrating the thermal efficiency of this vegetation cover compared to the present study's findings. Conversely, the elevated internal surface temperature (Figures 3B and 6B) in prototypes with fiber cement tiles without green roofing can be attributed to material darkening, which increases heat absorption from the external environment [64]. Additionally, Sampaio et al. [51], in a summer study conducted in southern Brazil, recorded lower fiber cement tile temperatures than those observed in this study without green roofing. This discrepancy can be explained by variations in climatic conditions, such as regional differences in temperature, humidity, and solar radiation, as well as the physical condition of the fiber cement roof tile. To optimize thermal performance in different climates, it can be beneficial to adjust or combine roofing materials, such as integrating green roofing or reflective coatings, to increase heat dissipation.

Beyond variations in internal surface temperature, it is crucial to relative humidity's impact on facility conditions. In the open prototype, relative humidity ranged between 40.8% and 74.3% (Figure 3C). These values are lower than those reported by Santos et al. [71], who documented monthly average relative humidity levels above 83.0% in poultry production environments in Sergipe. This difference can be attributed to their region's more humid coastal climate compared to the present experiment's location. Notably, prototypes with fiber cement roofing exhibited the lowest relative humidity levels during peak temperature periods between 12:00 P.M. and 3:00 P.M. (Figures 3C and 6C). This low relative humidity results from elevated air temperatures [72]. The green roof's substrate moisture retention likely contributed to increasing the relative humidity during peak heat periods, thereby improving environmental production conditions during these times. Relative humidity directly influences the thermal comfort and health of the animals. In the prototypes with green roofs (Figure 6C), humidity increased during the coldest hours due to evapotranspiration from the plants [70]. In prototypes closed on the sides, humidity exceeded 70%, regardless of the type of roof, which impairs the animals' transpiration, which can cause heat stress and respiratory diseases in the animals. Although green roofs improve the thermal conditions of the environment, it is necessary to balance the humidity when it is excessive. Ceramic tiles and fiber cement tiles have different characteristics: ceramic tiles dissipate heat better due to their greater thermal inertia [73], but they can retain more moisture, which favors condensation. Fiber cement tiles, on the other hand, absorb more heat, affecting the internal microclimate. Therefore, the combination of roofing materials and humidity control is fundamental to guaranteeing animal welfare. When well planned, the use of green roofs can promote a more balanced and healthier environment for animals.

The relative humidity values in the ceramic tile prototypes, both with and without green roofs, are particularly noteworthy (Figure 6C). Ceramic tiles' porosity enables environmental water absorption, contributing to their natural thermal performance by prolonging heat transmission [73]. However, this study's results indicate that ceramic tiles combined with green roofs promote higher relative humidity, which is potentially problematic in hot climates where it may compromise or impede air cooling processes. The disparity in mean values of relative humidity between the open and closed prototypes can be attributed to reduced ventilation in the internal environment and the effect of external airspeed, which removes heat from the surface of the tiles. Bollman et al. [15] reported similar findings, noting relative humidity changes due to water retention in green roof substrate, resulting in reduced internal building temperature. Despite variations, relative humidity values in green roof prototypes remained within the recommended range for poultry production (40.0% to 70.0%), as specified by Tinôco [52] and Ferreira [74]. Additionally, Bollman et al. [15] reported results similar to the present study (Figure 4), highlighting artificial reflective shadowing effects on relative humidity, with averages of 51.6% during the day and 72.7% at night in a tray system with a green roof in Corvallis, USA.

The results obtained in this study, when evaluating the thermal environment in prototypes with different types of roofing, made it possible to define thermal comfort indices (THI). In the prototypes with green roofs (Figure 4A), the average THI values provided optimal thermal comfort conditions, not only for the birds in the first week of life but throughout the production cycle, ranging from 72.0 to 80.0, as indicated by Silva et al. [50]. These values indicate that green roofs are effective in maintaining a comfortable environment for birds at different stages of growth [50]. In the prototypes without green roofs (Figure 7A), the THI approached the upper limit of this ideal range, with the fiber cement tile prototype exceeding the recommended range for broiler production [50]. This suggests that fiber cement roof tiles, because they do not have efficient thermal properties, contribute to an increase in internal temperature, compromising the birds' comfort, especially in warmer periods. This increase in temperature can lead to heat stress, negatively affecting the birds' performance [50]. These findings demonstrate that combining ceramic tiles with green roofing can enhance THI and consequently provide superior thermal comfort in the prototypes' internal environment. Passini et al. [75] identified average THI values similar to those found in the present study (Figures 4A and 7A), reaching an average of 77.0. Their study examined a broiler facility using reflective white paint on the external roof surface combined with artificial ventilation. However, the present study demonstrates that green roofing represents an effective alternative for promoting thermal comfort conditions; green roofs offer significant environmental and economic benefits, including reducing environmental impact through carbon sequestration and improving air quality. They also help retain moisture, reduce the need for irrigation, and reduce energy consumption by improving thermal efficiency and reducing reliance on artificial cooling systems. In economic terms, green roofs can reduce operational costs for ventilation and extend the life of the roof tiles, resulting in lower maintenance costs. This sustainable solution can, therefore, offer long-term benefits to producers, with potential government incentives and an appreciation of environmental commitment.

The thermal comfort analysis in reduced models demonstrates that green roofing can create a more thermally suitable environment for broilers, with results comparable to conventional methods such as combining white reflective paint with artificial ventilation [75]. However, Carneiro et al. [31], studying green roofs on reduced models of livestock facilities in Pernambuco state, observed average THI values around 26.0 for both peanut grass (Arachis repens) and *Zoysia japonica* grass. These values differ from the present study's findings, likely due to regional microclimatic variations.

These findings indicate that green roofing represents an effective construction technique with potential application in full-scale facilities. However, further detailed studies are needed to evaluate its practical implications for improving thermal conditions in animal production environments. As a sustainable alternative to conventional techniques, green roofing can significantly contribute to thermal comfort, reinforcing its value as an efficient and environmentally responsible solution.

Investigating green roof effects on BGHI during nighttime hours (6:00 P.M. to 06:00 A.M.) highlights the remarkable thermal insulation capacity of ceramic tiles with green covering, maintaining average BGHI values above other treatments (Figure 4B). The green roof's influence on BGHI is particularly evident during peak heat hours (9:00 A.M. to 3:00 P.M.), where average values in prototypes with this roofing reached 68.0 and 81.0 at 12:00 P.M. and 3:00 P.M. respectively, within the established thermal comfort range for broiler chickens, according to Oliveira et al. according to Oliveira et al. [52].

Furthermore, analysis of BGHI behavior in green ceramic tile prototypes without green roofing revealed values similar to those with metallic tiles and roofs (Figure 7B). Comparing this study's BGHI values with those documented by Oliveira et al. (2006) suggests that green roofing may enhance animal thermal comfort based on established BGHI values for adult-phase animals. Moreover, Sampaio et al. [51] observed an average BGHI value of approximately 80.2 in poultry house prototypes with lateral enclosures using ceramic tile roofing at 12:00 P.M., exceeding the animals' comfort range. This value closely mirrors the BGHI value obtained from identical tiles in the present investigation [73] Oliveira et al. (2006) [51] (Figure 7B).

Sampaio et al. [51], researching scaled-down animal production facility models, assert that vegetative roofing can induce cold stress, with an average BGHI of 70.0 varying by time of day and season. This contrasts with the higher mean BGHI values observed in green roof prototypes in the present study, which promote thermal comfort conditions (Figures 4B and 7B). Buffington et al. [35] establish that BGHI values up to 74.0 indicate comfort, while values above 84.0 signal emergency conditions for animal production, including poultry, cattle, and swine. The present study's results demonstrate that green roofs can provide a thermal environment conducive to animal requirements, potentially facilitating natural thermal conditioning within facilities [35].

Similarly, research by Sampaio et al. [51] reported that animal production facilities could achieve approximately 40.0% air temperature reduction during peak heat periods, consistent with this study's findings. However, green roofing led to even greater air temperature reductions (Figure 4C). The effectiveness of air temperature reduction varies with tile type when combined with green covering due to differing physical characteristics and thermal properties of materials. In enclosed prototypes (Figure 7C), green roofing also reduced average RTL values during peak heat periods, providing enhanced protection against incident solar radiation. The RTL values obtained in this study for prototypes with ceramic tiles and green roofs under both open and closed conditions (Figure 1) were comparable to those reported by Sampaio et al. [51]. They documented RTL values in Southern Brazil ranging from 406.7 to 479.2 W/m^2 for ceramic-tiled prototypes during winter and summer. Passini et al. [75] observed that white-painted roofing with artificial ventilation could reduce RTL by approximately 6.4 W/m^2. However, the current study demonstrated more substantial reductions in green-roofed prototypes. Notably, in enclosed prototypes with fiber cement tiles, the difference between those with and without green roofs was 74.0 W/m^2 at 12:00 P.M. and 69.0 W/m^2 at 3:00 P.M. (Figure 7C) [51,75]. The combination of ceramic tiles with a green roof has proven to be the most efficient in regulating the indoor microclimate, providing a significant improvement in thermal conditions. The combination of fiber cement tiles with a green roof, although less effective, still presents considerable benefits in mitigating heat. The use of green roofs combined with tiles with adequate thermal characteristics constitutes an effective strategy for optimizing thermal comfort and, consequently, animal welfare in production facilities, promoting a more stable environment that is favorable to productive performance.

Worker thermal environment assessment is essential, as it directly impacts health, well-being, and productivity [37]. This index provides a scientific basis for monitoring and adjusting environmental conditions within ideal comfort limits for specific activities, ensuring an adequate and safe workspace. In open prototypes, average HCI values without green roofing (Figure 4D) were higher during peak heat hours (9:00 A.M. to 3:00 P.M.), with notable peaks in fiber cement tile prototypes. However, during nighttime hours (6:00 P.M. to 6:00 A.M.), ceramic tiles with green covering demonstrated superior thermal insulation capacity, maintaining HCI above other treatments (Figure 4D).

Treatment disparities can be attributed to roofing materials' thermal properties [64]. Tiles without green covering protection tend to absorb more heat compared to covered tiles, which directly intercept solar radiation. This variation is crucial, particularly during peak heat hours when average HCI values approach 40.0 (Figure 7D), indicating uncomfortable worker conditions. Considering average HCI values in green-roofed prototypes, regardless of tile type (Table 2), this feature clearly provides comfortable worker conditions. This finding emphasizes green roofs' importance in promoting thermal comfort for both animals and facility workers.

5. Conclusions

Green roof implementation in animal production facilities has proven effective in enhancing the internal thermal environment, reducing temperature peaks, and promoting comfort—essential factors for animal welfare. This study's results demonstrate that green roofing, regardless of tile type, reduced indoor air temperature by up to 2.4 °C in open prototypes and 10.6 °C in closed prototypes during peak heat hours (3:00 P.M.). A significant reduction in the internal surface temperature of fiber cement tiles was also observed, reaching up to 24.0 °C in open prototypes and 27.0 °C in closed ones. The thermal comfort indexes (THI, BGHI, and RTL) also showed reductions, indicating green roofing's effectiveness in heat mitigation. The HCI for both types of green roof prototypes indicated comfortable conditions, complying with Ministry of Labor and Employment legislation.

Green roofs' natural thermal insulation reduces artificial cooling system requirements and represents a sustainable alternative for mitigating climate change effects on production facilities. Despite technical and financial challenges, including installation and maintenance costs, the environment and thermal comfort benefits justify increased investment and research to adapt this technology to agricultural applications, particularly in hot climates where thermal control is essential for animal production.

This study makes an innovative contribution by demonstrating, through experimental data, the thermal benefits of green roofs in reduced-scale poultry house prototypes. Green roof application in animal production facilities offers a viable solution for improving thermal conditions in these environments. The results provide valuable insights for future green roof implementation in the agricultural sector and can serve as a foundation for further research focused on enhancing thermal comfort and mitigating climate change impacts in rural animal production facilities.

Finally, carrying out analyses on reduced models of agricultural facilities is essential to assess the feasibility and performance of innovative technologies, such as green roofs, under controlled conditions before large-scale implementation. This approach enables a better understanding of thermal impacts, refinement of design parameters, and more precise estimation of costs and benefits, facilitating informed decisions that optimize thermal comfort and sustainability in animal production facilities.

Author Contributions: M.A.d.S.: Conceptualization, Data curation, Investigation, Methodology, Visualization, Writing—original draft, Writing—review and editing. F.C.d.S.: Conceptualization, Funding acquisition, Project administration, Resources, Supervision, Writing—review and editing. A.L.d.S.: Formal analysis, Software, Validation, Visualization. T.C.S.: Investigation, Methodology. C.P.O.: Investigation, Methodology. R.B.V.: Supervision, Writing—review and editing. F.d.C.B.: Conceptualization, Supervision, Writing—review and editing. I.d.F.F.T.: Resources, Supervision, Writing—review and editing. All authors have read and agreed to the published version of the manuscript.

Funding: This research was partially funded by the Coordination for the Improvement of Higher Education Personnel—Brazil (CAPES)—Financial Code 001.

Data Availability Statement: The data will be available upon request.

Acknowledgments: To the Research Center for Ambience and Agroindustrial Systems Engineering (AMBIAGRO) within the Department of Agricultural Engineering at the Federal University of Viçosa (UFV), to the Coordination for the Improvement of Higher Education Personnel, Brazil, (CAPES) with Financial Code 001, to the National Council for Scientific and Technological Development, Brazil (CNPq), and to the Minas Gerais Research Support Foundation, Brazil (FAPEMIG), we extend our sincere gratitude for the support provided.

Conflicts of Interest: The authors declare no conflicts of interest.

Symbols and Abbreviations

$W{\cdot}(m^2{\cdot}K^4)^{-1}$—watts per square meter per kelvin to the fourth power, $m{\cdot}s^{-1}$—meters per second, K—kelvin, %—percent, kPa—kilopascals, °C—degrees Celsius, m^2—square meters, Air Temperature—(AT), Relative Humidity—(RH), Temperature and Humidity Index—(THI), Black Globe Humidity Index—(BGHI), Human Comfort Index—(HCI), Radiant Thermal Load—(RTL), Ceramic Tile—(CT), Fiber Cement Tile—(FCT), Metal Tile—(MT), Ceramic Tile with Green Roof—(GCT), Fiber Cement Tile with Green Roof—(GFT), Metal Tile with Green Roof—(GMT), Crop Evapotranspiration—(ETc), Reference Evapotranspiration—(ET0).

References

1. ABPA. Associação Brasileira de Proteína Animal. 2023, pp. 1–75. Available online: https://abpa-br.org/ (accessed on 22 April 2024).
2. ABPA. Associação Brasileira de Proteína Animal. 2024, pp. 1–77. Available online: https://abpa-br.org/wp-content/uploads/2024/04/ABPA-Relatorio-Anual-2024_capa_frango.pdf (accessed on 22 April 2024).
3. Mascarenhas, N.M.H.; da Costa, A.N.L.; Pereira, M.L.L.; de Caldas, A.C.A.; Batista, L.F.; Andrade, E.L.G.; Mascarenhas, N.M.H.; da Costa, A.N.L.; Pereira, M.L.L.; de Caldas, A.C.A.; et al. Thermal conditioning in the broiler production: Challenges and possibilities. *J. Anim. Behav. Biometeorol.* **2020**, *6*, 52–55. [CrossRef]
4. Gonçalves, I.C.M.; Turco, S.H.N.; Neto, J.P.L.; Nascimento, J.W.B.D.; de Lima, V.L.A.; Borges, V.P. Thermal performance of aviary located in the semiarid region of Pernambuco based on computer simulation. *Rev. Bras. Eng. Agric. Ambient.* **2022**, *26*, 533–540. [CrossRef]
5. Mottet, A.; Tempio, G. Global poultry production: Current state and future outlook and challenges. *Worlds Poult. Sci. J.* **2017**, *73*, 245–256. [CrossRef]
6. Cheng, M.; McCarl, B.; Fei, C. Climate Change and Livestock Production: A Literature Review. *Atmosphere* **2022**, *13*, 140. [CrossRef]
7. Syarifuddin, H.; Sy, A.R.; Devitriano, D. CH4 Gas Mitigation Strategy with the Use of Interpretative Structural Modeling Method. In Proceedings of the 3rd Green Development International Conference (GDIC 2020); Atlantis Press: Amsterdam, The Netherlands, 2021; Volume 205, pp. 474–481. [CrossRef]
8. IPCC. *2014: Synthesis Report. Contribution of Working Groups I, II and III to the Fifth Assessment Report of the Intergovernmental Panel on Climate Change*; Core Writing Team, Pachauri, R.K., Meyer, L.A., Eds.; IPCC: Geneva, Switzerland, 2014.
9. Melo, T.V.; Furlan, R.L.; Milani, A.P.; Buzanskas, M.E.; de Moura, A.M.A.; Mota, D.A. Roof pitch and exposure and different roofing materials in reduced models of animal production facilities in the fall and winter. *Rev. Bras. Saúde Produção Anim.* **2015**, *16*, 658–666. [CrossRef]
10. Andric, I.; Kamal, A.; Al-Ghamdi, S.G. Efficiency of green roofs and green walls as climate change mitigation measures in extremely hot and dry climate: Case study of Qatar. *Energy Rep.* **2020**, *6*, 2476–2489. [CrossRef]
11. Baêta, F.C.; Souza, C.F. *Ambiência em Edificações Rurais: Conforto Animal*, 2nd ed.; UFV: Viçosa, Brazil, 2010.
12. Damasceno, F.A.; Schiassi, L.; Yanagi, T.; Osorio-Saraz, J.A.; Oliveira, J.L.-D. Evaluación térmica de tejas ecologicas en modelos físicos de galpones avicolas. *DYNA* **2016**, *83*, 114–119. [CrossRef]
13. Sampaio, C.A.P.; Terezo, R.F.; Motta, G.; Silva, L.M.C.; Júnior, I.V. Environmental thermal comfort of a reduced model using cross-laminated timber. *Eng. Agrícola* **2020**, *40*, 413–419. [CrossRef]
14. Shafique, M.; Kim, R.; Rafiq, M. Green roof benefits, opportunities and challenges—A review. *Renew. Sustain. Energy Rev.* **2018**, *90*, 757–773. [CrossRef]
15. Bollman, M.A.; DeSantis, G.E.; Waschmann, R.S.; Mayer, P.M. Effects of shading and composition on green roof media temperature and moisture. *J. Environ. Manag.* **2021**, *281*, 111882. [CrossRef]
16. La Roche, P.; Yeom, D.J.; Ponce, A. Passive cooling with a hybrid green roof for extreme climate. *Energy Build.* **2020**, *224*, 110243. [CrossRef]
17. Krebs, L.F.; Johansson, E. Influence of microclimate on the effect of green roofs in Southern Brazil—A study coupling outdoor and indoor thermal simulations. *Energy Build.* **2021**, *241*, 110963. [CrossRef]
18. Boafo, F.E.; Kim, J.T.; Kim, J.H. Evaluating the impact of green roof evapotranspiration on annual building energy performance. *Int. J. Green Energy* **2017**, *14*, 479–489. [CrossRef]
19. La Roche, P.; Berardi, U. Comfort and energy savings with active green roofs. *Energy Build.* **2014**, *82*, 492–504. [CrossRef]
20. Theodosiou, T. Green Roofs in Buildings: Thermal and Environmental Behaviour. *Adv. Build. Energy Res.* **2009**, *3*, 271–288. [CrossRef]

21. Lazzarin, R.M.; Castellotti, F.; Busato, F. Experimental measurements and numerical modelling of a green roof. *Energy Build.* **2005**, *37*, 1260–1267. [CrossRef]
22. Sailor, D.J.; Elley, T.B.; Gibson, M. Exploring the building energy impacts of green roof design decisions-a modeling study of buildings in four distinct climates. *J. Build. Phys.* **2012**, *35*, 372–391. [CrossRef]
23. Droz, A.G.; Coffman, R.R.; Blackwood, C.B. Plant diversity on green roofs in the wild: Testing practitioner and ecological predictions in three midwestern (USA) cities. *Urban For. Urban Green* **2021**, *60*, 127079. [CrossRef]
24. Sultana, M.N.; Akib, S.; Ashraf, M.A. Thermal comfort and runoff water quality performance on green roofs in tropical conditions. *Geol. Ecol. Landsc.* **2017**, *1*, 47–55. [CrossRef]
25. Yang, Y.; Davidson, C.I.; Zhang, J. Evaluation of thermal performance of green roofs via field measurements and hygrothermal simulations. *Energy Build.* **2021**, *237*, 110800. [CrossRef]
26. Houchmand, L.J.; Martí, M.M.; Gassó-Domingo, S. Photovoltaics and green roofs: Holistic analysis in built environments. *Renew. Sustain. Energy Rev.* **2025**, *207*, 114987. [CrossRef]
27. Yan, J.; Yang, P.; Wang, B.; Wu, S.; Zhao, M.; Zheng, X.; Wang, Z.; Zhang, Y.; Fan, C. Green Roof Systems for Rainwater and Sewage Treatment. *Water* **2024**, *16*, 2090. [CrossRef]
28. Jamei, E.; Chau, H.W.; Seyedmahmoudian, M.; Mekhilef, S.S.; Sami, F.A. Green roof and energy—Role of climate and design elements in hot and temperate climates. *Heliyon* **2023**, *9*, e15917. [CrossRef] [PubMed]
29. Kostadinović, D.; Jovanović, M.; Bakić, V.; Stepanić, N. Mitigation of urban particulate pollution using lightweight green roof system. *Energy Build.* **2023**, *293*, 113203. [CrossRef]
30. Talwar, P.; Verma, N.; Khatri, H.; Ahire, P.D.; Chaudhary, G.; Lindenberger, C.; Vivekanand, V. A systematic review of photovoltaic-green roof systems in different climatic conditions focusing on sustainable cities and societies. *Sustain. Cities Soc.* **2023**, *98*, 104813. [CrossRef]
31. Carneiro, T.A.; Guiselini, C.; Pandorfi, H.; Neto, J.P.L.; Loges, V.; de Souza, R.F.L. Condicionamento térmico primário de instalações rurais por meio de diferentes tipos de cobertura. *Rev. Bras. Eng. Agrícola Ambient.* **2015**, *19*, 1086–1092. [CrossRef]
32. Souza, M.A.; Sousa, F.C.; Baêta, F.C.; Vigoderis, R.B.; Zanetoni, H.H.R. Green roofs in animal production facilities—A review of strategies for estimating the carbon dioxide balance. *Renew. Sustain. Energy Rev.* **2024**, *189*, 114000. [CrossRef]
33. Thom, E.C. The Discomfort Index. *Weatherwise* **1959**, *12*, 57–61. [CrossRef]
34. Mader, T.L.; Davis, M.S.; Brown-Brandl, T. Environmental factors influencing heat stress in feedlot cattle. *J. Anim. Sci.* **2006**, *84*, 712–719. [CrossRef]
35. Buffington, D.E.; Collazo-Arocho, A.; Canton, G.H.; Pitt, D.; Thatcher, W.W.; Collier, R.J. Black globe-humidity index (BGHI) as comfort equation for dairy cows. *Elibrary. Asabe.Org.* **1981**, *24*, 711–0714. [CrossRef]
36. Rosenberg, V.S.B. *Microclimate: The Biological Environment*; John Wiley & Sons: New York, NY, USA, 1983.
37. Lei nº 6.514, de 22 de dezembro de 1977. Altera o Capítulo V, Título II, da Consolidação das Leis do Trabalho, relativa à Segurança e Medicina do Trabalho. Diário Oficial da União, Brasília, DF, 23 dez. 1977. Available online: https://legislacao.presidencia.gov.br/ (accessed on 22 April 2024).
38. Kelly, C.F.; Bond, T.E. Effectiveness of artificial shade materials. *Agric. Eng.* **1958**, *39*, 758–764.
39. Jentzsch, R.; Costa Baêta, F.; Tinôco, I.F.F.; Damasceno, F.A.; Cecon, P.R.; Saraz, J.A.O. Previsão de parâmetros térmicos ambientais dentro de modelos físicos em escala reduzida de alojamentos avícolas. *Interciencia* **2011**, *36*, 738–742. Available online: https://www.researchgate.net/publication/291497349 (accessed on 30 October 2024).
40. de Castro, A.C.; da Silva, I.J.O.; Nazareno, A.C.; Nunes, M.L.A.; Piedade, S.M.d.S. Thermal Efficiency of Different Coverage Materials in Reduced Models of Animal Husbandry Facilities: A Case Study. *Eng. Agrícola* **2017**, *37*, 403–413. [CrossRef]
41. Jentzsch, R.; Baêta, F.C.; Tinôco, I.F.F.; Damasceno, F.A.; Saraz, J.A.O. Parâmetros Arquitetônico-Ambientais para Construção e Testes em Modelos Reduzidos, Representativos de Galpões Avícolas, com Base em Similitude. *Rev. Eng. Agric.* **2013**, *21*, 19–30. [CrossRef]
42. Murphy, G. *Similitude in Engineering*; The Tonald Press Company: New York, NY, USA, 1950.
43. Bôas, R.L.V.; de Godoy, L.J.G.; Backes, C.; Santos, A.J.M.D.; Carribeiro, L.S. Sod production in Brazil. *Ornam. Hortic.* **2020**, *26*, 516–522. [CrossRef]
44. Martin, P.M. The potential of native grasses for use as managed turf. In Proceedings of the 4th International Crop Science Congress; CDROM: Brisbane, Australia, 2004; p. 13. Available online: https://www.cropscience.org.au (accessed on 25 October 2024).
45. Associação Nacional Grama Legal, Grama Esmeralda—Zoysia Japônica 2024. Available online: https://gramalegal.com/grama-esmeralda (accessed on 24 October 2024).
46. Rodríguez, J.; Vilela, K. Influence of the Types of Grass of Green Roofs for the Design of Thermal Comfort in Buildings. *J. Ecol. Eng.* **2022**, *23*, 223–229. [CrossRef]
47. Allen, R.; Pereira, L.; Raes, D.; Fao, M.S.; Rome, U. *Crop Evapotranspiration-Guidelines for Computing Crop Water Requirements-FAO Irrigation and Drainage Paper 56*; Food and Agriculture Organization of the United Nations: Rome, Italy, 1998.
48. INMET. 2022. Available online: https://tempo.inmet.gov.br/TabelaEstacoes/A510 (accessed on 19 July 2022).

49. Madeira, A.P.; Beneditto, D.; Doutor, S.S.; Zoologia, E. Necessidades hídricas das gramas batatais (Paspalum notatum Flüggé) e esmeralda (Zoysia Japônica Steud)estimadas por sensoriamento remoto. *Braz. J. Dev.* **2021**, *6*, 73015–73024.
50. Silva, E.T.d.; Leite, D.G.; Yuri, F.M.; Nery, F.d.S.G.; Rego, J.C.C.; Zanatta, R.d.A.; Santos, S.A.d.; Moura, V.V. Determinação do Índice de Temperatura e Umidade (ITU) para produção de aves na mesorregião metropolitana de Curitiba—PR. *Rev. Acadêmica Ciências Agrárias Ambient.* **2004**, *2*, 47–60.
51. Sampaio, C.A.P.; Cardoso, C.O.; Souza, G.P. Temperatura superficiais de telhas e sua relação com o ambiente térmico. *Eng. Agric.* **2011**, *31*, 1–11. [CrossRef]
52. Oliveira, R.F.M.; Donzele, J.L.; De Abreu, M.L.T.; Ferreira, R.A.; Vaz, R.G.M.V.; Cella, P.S. Effects of temperature and relative humidity on performance and yield of noble cuts of broilers from 1 to 49 days old. *Rev. Bras. Zootec.* **2006**, *35*, 797–803. [CrossRef]
53. Oliveira, C.P.; de Sousa, F.C.; Dallago, G.M.; Silva, J.R.; Campos, P.H.R.F.; Guimarães, M.C.d.C.; Baêta, F.d.C. Thermal Environment and Animal Comfort of Aviary Prototypes with Photovoltaic Solar Panel on the Roof. *Energies* **2023**, *16*, 2504. [CrossRef]
54. Brauer-Vigoderis, R.; Ferreira-Tinôco, I.D.F.; Pandorfi, H.; Bastos-Cordeiro, M.; De Souza-Júnior, J.P.; De Carvalho-Guimarães, M.C. Effect of heating systems in litter quality in broiler facilities in winter conditions. *Dyna* **2014**, *81*, 36. [CrossRef]
55. Carvalho, F.B.; Sartori, J.R.; Pezzato, A.C.; Fascina, V.B.; Castelo, P.G.; De Souza, I.M.G.P. Environmental temperature and broiler age on corn energy value. *Ciência Anim. Bras.* **2021**, *22*, e65526. [CrossRef]
56. Santos, W.M.M. *Índices de Conforto e Desconforto Térmico Humano segundo os Cenários Climáticos Do IPCC*; Congresso Brasileiro de Meteorologia: Belém, Brazil, 2010; p. 16.
57. de Souza, D.M.; Nery, J.T. O conforto térmico na perspectiva da Climatologia Geográfica. *GEOGRAFIA* **2012**, *21*, 65–83. [CrossRef]
58. Santos, G.; Miranda, B.; Diniz, F.R.; Da Silva, M.P. Avaliação e comparação do índice de conforto térmico humano entre as cidades de São Paulo (SP) e Bauru (SP). *J. Contrib.* **2017**, 1–5. [CrossRef]
59. TEAM, R.C.R. *R: A Language and Environment for Statistical Computing, 4.0.4 Los ed.*; R Foundation for Statistical Computing: Vienna, Austria, 2021.
60. Tinôco, I. Avicultura Industrial: Novos Conceitos de Materiais, Concepções e Técnicas Construtivas Disponíveis para Galpões Avícolas Brasileiros. *Rev. Bras. Cienc. Avic.* **2001**, *3*, 1–26. [CrossRef]
61. Collier, R.J.; Gebremedhin, K.G. Thermal Biology of Domestic Animals. *Annu. Rev. Anim. Biosci.* **2015**, *3*, 513–532. [CrossRef]
62. Pereira, F.; Ferreira, J.C.; Campos, A.T.; Ferreira, P.; Ferraz, P.; Bahuti, M.; Junior, T.Y.; Da Silva, J.P.; Ferreira, S.C. Dynamics of the Thermal Environment in Climate-Controlled Poultry Houses for Broiler Chickens. *Agriengineering* **2024**, *6*, 3891–3911. [CrossRef]
63. Michels, C.; Güths, S.; Marinoski, D.L.; Lamberts, R. Thermal performance and thermal resistance of fibre cement roof tiles: Experimental study. *Energy Build.* **2021**, *231*, 110569. [CrossRef]
64. Rivero, R. *Arquitetura e Clima: Acondicionamento Térmico Natural*, 2nd ed.; Rev. e Ampl.: Porto Alegre, RS, Brazil, 1986.
65. Verbeke, S.; Audenaert, A. Thermal inertia in buildings: A review of impacts across climate and building use. *Renew. Sustain. Energy Rev.* **2018**, *82*, 2300–2318. [CrossRef]
66. Barnabé, J.M.C.; Pandorfi, H.; de Almeida, G.L.P.; Guiselini, C.; Jacob, A.L. Temperatura superficial de materiais utilizados para cobertura individual de bezerreiros. *Rev. Bras. Eng. Agrícola Ambient.* **2014**, *18*, 545–550. [CrossRef]
67. Abreu, V.M.N.; de Abreu, P.G. Os desafios da ambiência sobre os sistemas de aves no Brasil. *Rev. Bras. Zootec.* **2011**, *40*, 1–14.
68. Cândido, M.G.; Tinôco, I.D.F.; Pinto, F.D.A.D.C.; Santos, N.T.; Roberti, R.P. Determination of thermal comfort zone for early-stage broilers. *Eng. Agrícola* **2016**, *36*, 760–767. [CrossRef]
69. Faggianelli, G.A.; Brun, A.; Wurtz, E.; Muselli, M. Natural cross ventilation in buildings on Mediterranean coastal zones. *Energy Build.* **2014**, *77*, 206–218. [CrossRef]
70. Pragati, S.; Priya, R.S.; Pradeepa, C.; Senthil, R. Simulation of the Energy Performance of a Building with Green Roofs and Green Walls in a Tropical Climate. *Sustainability* **2023**, *15*, 2006. [CrossRef]
71. Santos, G.D.B.; Sousa, I.F.D.; Brito, C.O.; Santos, V.D.S.; Barbosa, R.D.J.; Soares, C. Estudo bioclimático das regiões litorânea, agreste e semiárida do estado de Sergipe para a avicultura de corte e postura. *Ciência Rural.* **2014**, *44*, 123–128. [CrossRef]
72. Steadman, R.G. The Assessment of Sultriness. Part I: A Temperature-Humidity Index Based on Human Physiology and Clothing Science. *J. Appl. Meteorol. Climatol.* **1979**, *18*, 861–873. [CrossRef]
73. Bueno, A.D. Transferência de Calor e Umidade em Telhas: Simulação e Análise Experimental. Doctoral Dissertation, Universidade Federal de Santa Catarina, Florianópolis, Brazil, 1994.
74. Ferreira, R.A. *Maior Produção com Melhor Ambiente Para aves, Suínos e Bovinos*; Aprenda Fácil: Viçosa, Brazil, 2005.
75. Passini, R.; de Araújo, M.A.G.; Yasuda, V.M., Almeida, E.A. Intervenção ambiental na cobertura e ventilação artificial sobre índices de conforto para aves de corte. *Rev. Bras. Eng. Agrícola Ambient.* **2013**, *17*, 333–338. [CrossRef]

AgriEngineering

Article

Impact of Stocking Density on Welfare and Performance of Ross 708 and Cobb 700 Broilers

Shengyu Zhou [1], Tanner Thornton [1], Hao Gan [2], Tom Tabler [1], Mustafa Jaihuni [1], Xiaojuan Zhu [3] and Yang Zhao [1,*]

1 Department of Animal Science, The University of Tennessee, Knoxville, TN 37996, USA; szhou21@vols.utk.edu (S.Z.); tthorn15@vols.utk.edu (T.T.); gtabler@utk.edu (T.T.); mjaihuni@vols.utk.edu (M.J.)
2 Department of Biosystems Engineering & Soils Science, The University of Tennessee, Knoxville, TN 37996, USA; hgan1@utk.edu
3 Research Computing Support, Office of Innovative Technologies, The University of Tennessee, Knoxville, TN 37996, USA; xzhu8@utk.edu
* Correspondence: yzhao@utk.edu; Tel.: +1-865-974-6466

Abstract: Stocking density (SD) may affect broiler productivity and welfare. This study investigated the performance and welfare of Ross 708 and Cobb 700 broilers as affected by four SDs (27, 29, 32, and 44 kg/m^2) until day 56. A total of 432 birds per strain were used, with 10, 12, 14, and 18 birds per pen (1.1 m $\times$ 1.5 m), corresponding to the respective SDs. Each SD treatment comprised eight replicates. The target SD was determined based on the projected market weight of 4 kg at 56 days of age. The average body weight (BW), feed intake, and feed conversion ratio (FCR) were measured biweekly. Welfare indicators (four broilers per pen), including gait score, feather cleanliness, feather coverage, body temperature, and footpad condition, were evaluated on days 28 and 56. Tibia strength (two broilers per pen) was measured on day 56. The results show that the BW and FCR of both broiler strains were not affected by SD. For both strains, the male broilers exhibited greater bone strength compared to that of the female broilers (129.06 lbf M vs. 91.70 lbf F for Ross, and 130.86 lbf M vs. 117.40 lbf F for Cobb), but the influence of SD on bone strength was found to be significant only for the Ross male broilers. Most welfare indicators were not affected by the SD on days 28 and 56 for either broiler strain, except for feather cleanliness in Ross broilers and footpad in Cobb broilers on day 56, which improved at lower SDs. Strong age and sex effects on the welfare indicators were also identified for both broiler strains. It was concluded that the SD is not a significant factor for broiler productivity, and it has a minor influence on broiler welfare compared to those of age and sex.

Keywords: broiler; stocking density; production performance; welfare

1. Introduction

The broiler industry is crucial in meeting the protein needs of the U.S. In 2022, per capita consumption of broiler products in the United States reached 44.1 kg, nearly double the consumption of beef or pork [1]. To meet this high demand, 20.6 billion kg of live broilers were produced in 2022, with a total market value exceeding USD 66 billion [1]. Additionally, the broiler industry significantly contributed to the U.S. economy, generating over 1.5 million jobs and adding USD 417 billion in economic impact [2].

The broiler industry has made significant reductions in the environmental footprint [3] through improvements in productivity and efficiency over the years [4], providing a vital source of affordable yet high-quality animal protein. Modern commercial broiler strains achieve nearly four times the body weight (BW) at day 56 compared to strains from six decades ago, along with a significant 40% improvement in feed conversion ratio (FCR) [5]. However, rapid growth rates and intensive commercial production systems have raised growing public concerns regarding animal welfare [6,7].

Animal welfare is guided by the "five freedoms" [8], which Mellor [9] later refined to include good nutrition, a healthy environment, good health, appropriate behavior, and a positive mental experience. Fast growth and efficiency-oriented feeding and management procedures have led to welfare issues in modern broilers [10]. Welfare issues in broilers are multifaceted, including lameness, footpad dermatitis, metabolic disorders [10], poor environmental conditions [11], etc. Some major welfare concerns are related to stocking density (SD). The SD may affect the litter and air quality, which are major welfare, environmental, and management concerns at commercial broiler farms. High SD can restrict normal behavior in birds and lead to increased moisture in litter due to higher manure production per unit area. High moisture in the litter may foster microbial growth and increase ammonia levels in broiler houses [12], consequently posing the risk of contact footpad dermatitis. Furthermore, high SD restricts heat transfer from the litter to the surrounding space, reducing the effectiveness of ventilation systems in managing heat stress [10].

The current recommended SD in the U.S. varies. The National Chicken Council (NCC) Broiler Welfare Guidelines, widely recognized as the standard for broiler welfare in the U.S. [13], suggest a range of 32 to 44 kg/m^2. This range is based on the birds' final market weight and takes into account the space needed for broilers to express normal behaviors [14]. In comparison, animal welfare programs like the Global Animal Partnership (GAP) recommended lower SDs, from 27 to 29 kg/m^2, to accommodate natural bird behaviors. According to Mckenna [15], only 600 farms (2.4%) out of 25,000 in the U.S. [16] have adopted the GAP SDs for broiler production. The low adoption rate is possibly a result of high production costs. In addition, a lower SD may also be associated with an increase in land use and a larger environmental footprint [17]. Although a lower SD has the potential to enhance welfare, it is essential to balance sustainable production, land use, environmental impact, and animal welfare.

The market share of commercial broilers is mainly dominated by the Ross and Cobb strains in the United States [18]. The production responses of these two strains to SD are different. For Ross broilers raised to approximately 42 days of age, an SD of 13–15 birds/m^2 (equivalent to 32.5–37.5 kg/m^2, based on an average weight of 2.5 kg at 42 days) was recommended for achieving higher BWs [19–21] and a lower FCR [22–24]. In comparison, a density of 10 birds/m^2 (28.5 kg/m^2, calculated based on an average weight of 2.85 kg at 42 days) for raising Cobb broilers is suggested for maximizing BW [25–27]. However, an SD of 17 birds/m^2 for Cobb broilers provided the best economic return of USD 3.98/m^2, despite a slightly lower BW and FCR [27]. Currently, there is a lack of research comparing the NCC- and GAP-recommended SDs under similar production conditions. As such, investigating the impacts of NCC- and GAP-recommended SDs on major broiler strains under similar production conditions is warranted.

The objective of this study was to evaluate the performance and welfare of Ross 708 and Cobb 700 broilers under different SDs. Precision livestock farming (PLF) technologies were utilized for an objective assessment of feather coverage through temperature analysis.

2. Materials and Methods

2.1. Birds, Diets, and Management

A total of 432 one-day-old broilers from each strain (obtained from a local commercial hatchery) were raised under four SDs. The initial weight of the Ross 708 birds was about 38 g, and that of the Cobb 700 birds was about 36 g. Identical pens (1.1 m × 1.5 m) were stocked with 10, 12, 14, or 18 birds, and each SD treatment had eight replicate pens. The target SD was determined based on the projected market weight of 4 kg at 56 days of age. Due to the limited size of the experimental room, we conducted the trial with Ross 708 broilers in September-October 2022 and the trial with Cobb 700 broilers in January-February 2023. Each pen was equipped with one 36 cm diameter tube feeder and three nipple drinkers. The feeder space and the number of drinkers were sufficient for all treatments, based on the NCC recommendation [14]. The broilers were feather-sexed (for Ross birds) or vent-sexed (for Cobb birds) and distributed evenly in experimental pens at a 1:1 male-to-female ratio. Environment management, including light intensity and temperature, in the experimental room was adjusted according to the Ross 708 [28] and the Cobb 700 [29] broiler management guidelines by bird age. At the study's commencement, each pen was bedded with topsoil (Farmer Green topsoil, 0.14 cubic feet per pen) and covered with black mulch (Earthgro black wood shredded mulch, 0.04 cubic feet per pen), which remained unchanged throughout the study period. Every two weeks, the total weight of all birds in each experimental pen was determined. Feed intake was also measured biweekly to calculate the FCR (feed intake/body weight gain). The empty feed bucket was weighed initially. Feed was added daily, and the amount provided was recorded. Every two weeks, the bucket, with any remaining feed, was weighed, and the total feed intake for the pen over that period was calculated by subtracting the final feed weight from the total weight of the feed added during the two weeks. The broilers were raised over a 56-day production cycle.

2.2. Welfare and Behavior Measurement

Broiler welfare was assessed manually based on feather cleanliness, footpad dermatitis, gait score, and bone strength according to the methods of Zhou et al. [30], and the Welfare Quality® assessment protocol [31]. Scores for each indicator reflected varying degrees of welfare, with lower scores representing better conditions. For feather cleanliness, the scores ranged from 0 (clean) to 3 (severely soiled), and for footpad dermatitis, the scale extended from 0 (no lesions) to 4 (severe lesions). Gait scores ranged from 0 (normal mobility) to 5 (unable to walk), with gait scores of 3 and above signaling a welfare concern [32].

Precision agriculture technologies, including thermography and image processing [33], were utilized to measure the bare-skin ratio and surface body temperature (Figure 1. For Ross and Cobb broilers at 28 days of age, a threshold of 33.5 °C was used to calculate the bare-skin ratio. At 56 days of age, the threshold was set at 35 °C for the calculation. The back and belly parts correspond to the back and breast regions illustrated by Zhao et al. [33].

(**a**) (**b**)

Figure 1. These thermal images depict a Cobb broiler (SD-44 treatment, Pen 31, Bird #5) on day 56. (**a**) A thermal image of the broiler's back, with the white polygon outlining the back area. Using a 35 °C threshold, the average temperature of the area was calculated to be 25.4 °C, and no bare skin was detected (0% bare-skin ratio). (**b**) A thermal image of the broiler's belly, with the white polygon outlining the belly area. With the same 35 °C threshold, the average temperature was measured at 34.2 °C, and the bare-skin ratio was 55.5%. The red dots in the figure indicate the highest temperature points within the selected range, while the blue dots represent the lowest temperature points. These measurements were automatically determined using the "FLIR Thermal Studio Pro" software. Any overlapping content in the figure was generated automatically by the software's measurement tools and can be disregarded.

At 28 days of age, four birds (two males and two females) per pen were randomly selected for assessments of feather cleanliness, footpad dermatitis, and gait score. At 56 days of age, four birds (two males and two females) were randomly selected for the same assessments. During measurements, the selected birds were temporarily removed from their pens. A thermal camera (T865, Teledyne FLIR, Wilsonville, OR) was used to measure the back and belly surface temperatures, allowing for the calculation of the ratio of bare skin to feather coverage. Additionally, two broilers (one male and one female) per pen were euthanized using the carbon dioxide method for tibia-breaking strength assessment at the trial's conclusion. The left tibia-breaking strength was measured using a three-point bending method with an MTS Alliance RT/30 apparatus (MTS Systems Corporation, Eden Prairie, MN), maintaining a 4 cm distance between the two supporting points [30]. All procedures followed the guidelines in the Guide for the Care and Use of Agricultural Animals in Research and Teaching [34] and were approved by the University of Tennessee's Institutional Animal Care and Use Committee (IACUC Protocol #2876-1221).

2.3. Statistical Analysis

All statistical analyses were performed using JMP, version 16.0.0 (SAS Institute, Cary, NC, USA). A randomized complete block design was employed with SDs of 27, 29, 32, and 44 kg/m^2, each having eight replications. Broilers were randomly selected at different ages, meaning that the individuals selected at days 28 and 56 were not necessarily the same. A mixed analysis of variance (ANOVA) was used to assess the impact of SD on continuous variables (bone-breaking strength, average temperature, and bare-skin ratio), with pens as the random block effect. All scored variables, including gait score, footpad dermatitis, feather coverage, and feather cleanliness, were analyzed using a mixed ANOVA on ranks with pens as the random block effect to assess the effect of SD. Tukey's HSD test was used for post hoc pairwise comparisons with significance set at the 0.05 level, following the calculation of least squares means. The normality of residuals was assessed using the Shapiro–Wilk normality test. Additionally, the Kruskal–Wallis test was utilized to evaluate differences between ages (days 28 and 56) for scoring and continuous variables, with significance set at $p < 0.05$.

3. Results

The intention of this study was not to compare the Ross 708 and Cobb 700 strains but rather to evaluate the responses of each strain independently. Despite our efforts to maintain similar personnel, management, and environmental conditions for both strains, some variation in management was inevitable. Furthermore, the breeder hen age plays a crucial role in the performance of their offspring. Since we only employed one batch of chicks from each strain, without considering their parent age, it would neither be feasible nor appropriate to compare them due to the lack of replication.

3.1. Effects of SD on BW and FCR

Table 1 shows the effects of SD on BW and FCR for Ross and Cobb broilers. Stocking density did not significantly affect the BW and FCR of Ross 708 and Cobb 700 broilers. At 28 days, the Ross 708 broilers had an average weight of 1.01 kg ($p = 0.57$) and an FCR of 1.81 ($p = 0.57$). At 56 days, the Ross birds reached an average weight of 3.59 kg ($p = 0.57$), with an FCR of 1.90 ($p = 0.57$). The Cobb broilers weighed 1.42 kg at day 28 ($p = 0.69$), with an FCR of 1.62 ($p = 0.15$). By day 56, their average BW had increased to 4.31 kg ($p = 0.15$). Due to accidentally missing data, it was not possible to calculate FCR on day 56.

Table 1. Effect of stocking density (SD, kg/m^2) on the body weight (BW, kg) and feed conversion ratio (FCR) of Ross 708 and Cobb 700 broilers.

	Effect of SD on the BW and FCR of Ross 708						
BW (kg) or FCR	SD 27	SD 29	SD 32	SD 44	Average	Std Error	*p*-value
BW (Day 14)	0.33	0.34	0.33	0.33	0.33	<0.01	0.35
BW (Day 28)	1.00	1.10	0.97	0.97	1.01	0.02	0.48
BW (Day 42)	2.10	2.06	2.02	2.06	2.06	0.04	0.61
BW (Day 56)	3.66	3.56	3.54	3.60	3.59	0.07	0.57
FCR (Day 0–28)	1.83	1.77	1.75	1.88	1.81	0.03	0.06
FCR (Day 0–56)	1.92	1.93	1.89	1.87	1.90	0.03	0.57
	Effect of SD on the BW and FCR of Cobb 700						
BW (kg) or FCR	SD 27	SD 29	SD 32	SD 44	Average	Std Error	*p*-value
BW (Day 14)	0.39	0.40	0.40	0.39	0.40	0.01	0.40
BW (Day 28)	1.44	1.40	1.42	1.41	1.42	0.02	0.69
BW (Day 42)	2.80	2.80	2.77	2.67	2.76	0.05	0.20
BW (Day 56)	4.38	4.35	4.35	4.15	4.31	0.08	0.15
FCR (Day 0–28)	1.64	1.66	1.56	1.63	1.62	0.03	0.15

3.2. *Effects of SD, Sex, and SD by Sex Interactions on Bone-Breaking Strength*

Tables 2 and 3 show the effects of SD, sex, and the SD by sex interactions on bone-breaking strength for Ross and Cobb broilers. Reducing the SD from 32 to 27 reduced the bone-breaking strength of male Ross broilers from 142.95 to 101.82 lbf ($p = 0.01$), while the bone strength of female Ross birds was not affected by SD, with an average bone-breaking strength of 91.70 lbf. There was no interaction effect of SD and sex on Cobb broilers ($p = 0.79$). The bone strength of Ross (average 110.38 lbf, $p = 0.14$) and Cobb (average 124.12 lbf, $p = 0.11$) broilers was not affected by SD. The bone strength of male broilers was higher than that of females for both the Ross (37.36 lbf higher, $p < 0.01$) and Cobb (13.46 lbf higher, $p < 0.01$) birds.

Table 2. Effects of stocking density (SD, kg/m^2), sex, and SD by sex interaction on Ross 708 broiler's bone-breaking strength (lbf).

	Effect of SD on bone strength of Ross						
SD 27	SD 29		SD 32	SD 44	Average	Std Error	*p*-value
99.19	114.63		116.33	111.36	110.38	5.65	0.14
	Effect of sex on bone strength of Ross						
	Male		Female		Std Error		*p*-value
	129.06 ^{a}		91.70 ^{b}		3.89		<0.01
	SD by sex interaction effects of Ross						
	SD 27	SD 29	SD 32	SD 44	Std Error		*p*-value
Male	101.82 ^{bc}	135.49 ^{ab}	142.95 ^{a}	135.99 ^{ab}	7.72		0.01
Female	96.56 ^{c}	93.78 ^{c}	89.72 ^{c}	86.73 ^{c}			

$^{a, b, c}$ means in the same row with no common superscripts differ ($p < 0.05$).

Table 3. Effects of stocking density (SD, kg/m^2), sex, and SD by sex interaction on Cobb 700 broiler's bone-breaking strength (lbf).

	Effect of SD on bone strength of Cobb						
SD 27	SD 29		SD 32	SD 44	Average	Std Error	*p*-value
120.27	131.98		116.07	128.16	124.12	5.02	0.11
	Effect of sex on bone strength of Cobb						
	Male		Female		Std Error		*p*-value
	130.86 ^{a}		117.40 ^{b}		3.55		<0.01

Table 3. *Cont.*

	SD by sex interaction effects of Cobb					
	SD 27	SD 29	SD 32	SD 44	Std Error	*p*-value
Male	125.64	138.58	120.13	139.11	7.09	0.79
Female	114.89	125.40	112.02	117.21		

$^{a, b}$ means in the same row with no common superscripts differ ($p < 0.05$).

3.3. Effects of SD on Gait Score, Footpad Dermatitis, Feather Cleanliness, Back and Belly Temperature, and Bare-Skin Ratio

Tables 4 and 5 show the effects of SD on gait score, footpad dermatitis, feather cleanliness, back and belly temperature, and bare-skin ratio for Ross and Cobb broilers. The welfare values in these tables represent the average welfare scores for birds under each treatment.

Table 4. Effects of stocking density (SD, kg/m^2) on the gait score, footpad dermatitis, feather cleanliness, back and belly body temperature (°C), and bare-skin ratio (%) of Ross 708 broilers. Scores for feather cleanliness, gait score, and footpad dermatitis were determined following the Welfare Quality® (2009) assessment protocol.

Measurement Items	Ross 708						
On day 28	SD 27	SD 29	SD 32	SD 44	Average	Std Error	*p*-value
Gait score	0.31	0.34	0.32	0.25	0.31	0.09	0.93
Footpad	1.84	2.16	1.91	2.16	1.84	0.25	0.77
Feather cleanliness	0.06	0.00	0.00	0.03	0.06	0.03	0.35
Back temperature	31.96	31.88	32.68	32.46	32.25	0.37	0.36
Belly temperature	36.62	36.99	37.21	36.91	36.93	0.18	0.17
Back-skin ratio	19.73	19.27	27.25	22.50	22.19	3.15	0.28
Belly-skin ratio	64.61	65.77	69.08	64.69	66.04	1.89	0.32
On day 56	SD 27	SD 29	SD 32	SD 44	Average	Std Error	*p*-value
Gait score	1.56	1.16	1.50	1.22	1.36	0.16	0.38
Footpad	0.94	1.09	1.22	1.50	1.19	0.15	0.08
Feather cleanliness	1.34 ^{b}	1.41 ^{b}	1.59 ^{b}	2.19 ^{a}	1.63	0.15	<0.01
Back temperature	25.03	24.80	25.15	24.89	24.97	0.28	0.82
Belly temperature	31.75	31.82	32.23	31.82	31.91	0.24	0.46
Back-skin ratio	2.35	3.29	3.49	3.30	3.11	0.84	0.77
Belly-skin ratio	54.49	55.66	59.43	56.99	56.64	1.98	0.35

$^{a, b}$ means in the same row with no common superscripts differ ($p < 0.05$).

For Ross 708 broilers, the gait score (0.31, p = 0.93) and footpad dermatitis (1.84, p = 0.77) on day 28 and the gait score (1.36, p = 0.38) and footpad dermatitis (1.19, p = 0.08) on day 56 were not affected by SD. Lowering the SD improved feather cleanliness for the SD-27 (1.34), SD-29 (1.41), and SD-32 (1.59) groups compared with those of the SD-44 (2.19) birds on day 56 ($p < 0.01$). On day 28, SD did not affect back or belly temperature or bare-skin ratio. The belly temperature was 4.68 °C higher and the bare-skin ratio was 43.85% higher the scores for the back. Similarly, on day 56, SD had no significant effect on these parameters, with the belly temperature 6.94 °C higher and the bare-skin ratio 53.53% higher than the scores for the back.

For Cobb 700 broilers, SD affected the footpad on day 56 (p = 0.04), with birds at 27 kg/m^2 (1.22) displaying healthier footpads than those at 44 kg/m^2 (2.27). The gait scores, feather cleanliness, back and belly temperature, and bare-skin ratio of the Cobb birds were unaffected by SD on either day 28 or day 56. On day 28, the belly temperature was 3.88 °C higher, and the belly bare-skin ratio was 40.3% greater than that for the back.

Similarly, on day 56, the belly temperature exceeded the back temperature by 7.17 °C, and the belly bare-skin ratio was 56.43% higher than that for the back.

Table 5. Effects of stocking density (SD, kg/m^2) on the gait score, footpad dermatitis, feather cleanliness, back and belly body temperature (°C), and bare-skin ratio (%) of Cobb 700 broilers.

Measurement Items	**Cobb 700**						
On day 28	SD 27	SD 29	SD 32	SD 44	Average	Std Error	*p*-value
Gait score	0.41	0.19	0.56	0.47	0.41	0.12	0.18
Footpad	1.16	1.47	1.53	1.56	1.43	0.26	0.62
Feather cleanliness	0.72	0.44	0.47	0.63	0.56	0.13	0.47
Back temperature	31.29	30.28	31.23	30.63	30.86	0.38	0.20
Belly temperature	35.12	34.52	34.66	34.67	34.74	0.28	0.45
Back-skin ratio	28.91	22.93	30.64	26.80	27.32	3.35	0.41
Belly-skin ratio	71.61	66.85	65.93	66.10	67.62	2.37	0.30
On day 56	SD 27	SD 29	SD 32	SD 44	Average	Std Error	*p*-value
Gait score	1.97	2.00	2.00	1.87	1.96	0.13	0.89
Footpad	1.22 ^{b}	1.81 ^{ab}	1.86 ^{ab}	2.27 ^{a}	1.79	0.24	0.04
Feather cleanliness	2.09	2.44	2.22	2.45	2.30	0.18	0.57
Back temperature	24.55	24.48	24.62	24.33	24.50	0.24	0.84
Belly temperature	32.00	31.48	31.62	31.58	31.67	0.28	0.57
Back-skin ratio	0.80	0.93	0.96	0.48	0.80	0.25	0.55
Belly-skin ratio	59.03	55.69	56.97	57.21	57.23	2.82	0.87

$^{a, b}$ means in the same row with no common superscripts differ ($p < 0.05$).

3.4. Effect of Age on Gait Score, Footpad Dermatitis, Feather Cleanliness, Back and Belly Temperature, and Bare-Skin Ratio

Table 6 highlights the significant effects of age on broiler welfare indicators. From day 28 to day 56, the footpad health in the Ross birds improved by 0.78, while in the Cobb birds, it declined by 0.36 (both $p < 0.01$). The gait scores worsened, and feather cleanliness declined for both the Ross and Cobb broilers over this period. However, the average back and belly temperatures, along with bare-skin ratios, decreased, indicating improved feather coverage with age (all $p < 0.01$).

Table 6. Effects of age on the gait score, footpad dermatitis, feather cleanliness, back and belly body temperature (°C), and bare-skin ratio (%) of Ross 708 and Cobb 700 broilers.

Measurement Items	Ross 708			
	Day 28	Day 56	Std Error	*p*-value
Gait score	0.30 ^{b}	1.36 ^{a}	0.16	<0.01
Footpad	1.97 ^{a}	1.19 ^{b}	0.10	<0.01
Feather cleanliness	0.02 ^{b}	1.63 ^{a}	1.27	<0.01
Back temperature	32.24 ^{a}	24.97 ^{b}	0.95	<0.01
Belly temperature	36.93 ^{a}	31.90 ^{b}	0.07	<0.01
Back-skin ratio	22.19 ^{a}	3.11 ^{b}	0.11	<0.01
Belly-skin ratio	66.04 ^{a}	56.64 ^{b}	0.05	<0.01
	Day 28	Day 56	Std Error	*p*-value
Gait score	0.41 ^{b}	1.95 ^{a}	0.06	<0.01
Footpad	1.43 ^{b}	1.79 ^{a}	0.12	<0.01
Feather cleanliness	0.56 ^{b}	2.30 ^{a}	0.08	<0.01
Back temperature	30.86 ^{a}	24.49 ^{b}	0.15	<0.01
Belly temperature	27.32 ^{a}	0.81 ^{b}	1.20	<0.01

Table 6. *Cont.*

Measurement Items	Ross 708			
	Day 28	Day 56	Std Error	*p*-value
Back-skin ratio	34.74 [a]	31.67 [b]	0.13	<0.01
Belly-skin ratio	67.62 [a]	57.24 [b]	1.13	<0.01

[a, b] means in the same row with no common superscripts differ ($p < 0.05$).

3.5. *Effect of Sex on Gait Score, Footpad Dermatitis, Feather Cleanliness, Back and Belly Temperature, and Bare-Skin Ratio*

Table 7 illustrates the effects of sex on welfare indicators in Ross 708 and Cobb 700 broilers. At 28 days of age, no significant differences were observed between males and females for any welfare parameters in either strain. By 56 days of age, however, several significant differences emerged. In the Ross 708 broilers, females exhibited lower back temperatures (24.58 °C vs. 25.35 °C; $p < 0.01$) and a reduced back-skin ratio (1.70% vs. 4.52%; $p < 0.01$) compared to those of males. No significant differences were found in regards to gait score, footpad condition, feather cleanliness, belly temperature, or belly-skin ratio for the Ross 708 broilers. In the Cobb 700 broilers at 56 days of age, females showed improved gait scores (1.78 vs. 2.14; $p = 0.01$), healthier footpad scores (1.63 vs. 1.95; $p = 0.04$), and cleaner feathers (2.21 vs. 2.39; $p = 0.04$) compared to those of males. However, no significant differences were observed between males and females for back temperature, belly temperature, back-skin ratio, or belly-skin ratio in Cobb 700 broilers.

Table 7. Effects of sex on the gait score, footpad dermatitis, feather cleanliness, back and belly body temperature (°C), and bare-skin ratio (%) of Ross 708 and Cobb 700 broilers.

Measurement Items	**Ross 708**				**Cobb 700**			
On day 28	Male	Female	Std Error	*p*-value	Male	Female	Std Error	*p*-value
Gait score	0.33	0.28	0.06	0.58	0.38	0.44	0.09	0.37
Footpad	2.01	1.92	0.16	0.65	1.27	1.59	0.16	0.07
Feather cleanliness	0.02	0.03	0.02	0.58	0.63	0.50	0.08	0.22
Back temperature	32.04	32.45	0.28	0.34	30.65	31.06	0.08	0.30
Belly temperature	36.96	36.91	0.12	0.74	30.65	31.06	0.27	0.87
Back-skin ratio	20.31	24.07	2.53	0.34	25.26	29.38	0.27	0.23
Belly-skin ratio	67.07	65.00	1.32	0.26	34.76	34.73	0.17	0.72
On day 56	Male	Female	Std Error	*p*-value	Male	Female	Std Error	*p*-value
Gait score	1.45	1.27	0.12	0.38	2.14 [a]	1.78 [b]	0.09	0.01
Footpad	1.17	1.20	0.12	0.78	1.95 [a]	1.63 [b]	0.15	0.04
Feather cleanliness	1.63	1.64	0.09	0.82	2.39 [a]	2.21 [b]	0.10	0.04
Back temperature	25.35 [a]	24.58 [b]	0.18	<0.01	24.56	24.42	0.14	0.43
Belly temperature	31.94	31.87	0.15	0.73	31.50	31.83	0.17	0.13
Back-skin ratio	4.52 [a]	1.70 [b]	0.60	<0.01	0.97	0.62	0.18	0.19
Belly-skin ratio	56.71	56.57	1.30	0.93	55.40	59.05	1.70	0.06

[a, b] means in the same row with no common superscripts differ ($p < 0.05$).

4. Discussion

4.1. *Effects of Stocking Density on Body Weight and FCR*

The SD had no significant effect on the BW or FCR of the Ross and Cobb broilers, with the final BW averaging 3.59 kg for the Ross and 4.31 kg for the Cobb birds. These results align with the findings by Wang et al. [35], who reported no significant influence of SD on broiler performance. Weimer et al. [36] also found that the broiler's BW (2.714 kg) and FCR (1.55) on day 42 were unaffected by SD, specifically comparing with 29 kg/m^2 and 37 kg/m^2. These findings suggest that precise control of SD can allow poultry producers to optimize space usage without compromising growth performance. However, contrasting studies have reported that higher SDs can negatively impact broiler BW and feed intake under specific conditions [25,27,37–39]. Son et al. [40] demonstrated that increased densities (23–26 birds/m^2) under heat stress conditions led to reduced BW and increased oxidative stress markers. Consistent with this, Goo et al. [38] found that a high SDs combined with heat stress significantly impaired broiler growth performance. These findings highlight the importance of considering environmental factors, such as temperature and ventilation, alongside SD. Differences in results across studies may stem from variations in feeding strategies, rearing environments, or genetic strains. As noted by Ekstrand and Carpenter [41], the impact of SD on broiler performance is strongly influenced by environmental conditions, such as litter quality and ventilation. Therefore, a holistic management approach addressing both density and environmental variables is essential to mitigate potential adverse effects. Our findings reinforce the efficacy of SD recommendations, such as the NCC's recommended 44 kg/m^2 [14]. Adherence to such guidelines supports consistent production outcomes, while maintaining bird welfare. This approach allows producers to optimize economic returns without negatively affecting animal welfare or growth performance, making it a practical strategy for sustainable poultry production.

4.2. *Effects of SD, Sex, and SD by Sex Interactions on Bone-Breaking Strength*

Generally speaking, broilers raised at higher densities exhibit lower bone-breaking strength compared to those raised at lower densities [20,42–44]. The limited space affects their movement and bone development, leading to weaker bones and higher susceptibility to leg issues [44,45]. However, lowering the SD was sometimes related to compromised leg health. Tablante et al. [46] reported that reducing the SD from 20 to 10 birds/pen increased the incidence of tibial dyschondroplasia by 10.6%. Buijs et al. [47] found an unexpected latency-to-lie improvement (related to better leg health and walking ability) in birds raised under an SD of 47 kg/m^2 compared to birds raised at an SD ranging from 23 to 41 kg/m^2. The significance reported in this study was for males; reducing the SD from 32 to 27 reduced the bone-breaking strength of male Ross broilers, while the bone strength of female Ross birds was not affected by the reduction of SD. Although broilers raised at a lower SD had more space to move around, the lack of competition for feeding between male broilers due to the decreased SD may be the reason for the reduced bone-breaking strength.

The bone strength of Cobb broilers was not affected by SD. This is inconsistent with the previous report that increasing SD will reduce bone-breaking strength at around 45 days of age [20,30,43]. The fact that broilers were selected at different ages for measuring bone strength may explain this discrepancy. The bone-breaking strength in this experiment was measured at 56 days of age; the additional 11 days may have allowed bones to become stronger, eliminating the potential impact of SD. The bone-breaking strength of the Cobb birds at 56 days of age in this experiment was 124.12 lbf, which was higher than the 91.25 lbf of the Cobb at 45 days reported by Zhou et al. [30]. The bone strength of male broilers was higher than that of females for both the Ross and Cobb birds. This is in line with our expectations and consistent with previously reported results [48–50]. Mönch et al. [51] suggested that hormonal differences, which affect skeletal development, might explain the variance in bone strength between male and female broilers.

4.3. *Effects of SD on Gait Score, Footpad Dermatitis, Feather Cleanliness, Back and Belly Temperature, and Bare-Skin Ratio*

Reducing SD improved Ross's feather cleanliness at 56 days. Feathers serve as a protective layer, maintaining body temperature and shielding birds from physical damage [52]. Clean feathers are a critical welfare indicator, and improving feather cleanliness through optimized SD could enhance overall welfare. A higher SD, such as SD-44 in Ross broilers, compromised feather cleanliness at 56 days, likely due to overcrowding. Overcrowding increases feeding, drinking, and excretion activities, leading to elevated moisture and ammonia levels in litter, which are known to negatively affect feather and footpad health [53,54]. Adjusting the SD during specific growth phases may help maintain better feather and litter quality, minimizing welfare issues.

Lowering the SD improved the Cobb's footpad health at 56 days, as a higher SD was associated with wetter litter, compromising footpad conditions [53]. Integrating litter management systems to control moisture and ammonia levels could enhance footpad health outcomes, reducing potential welfare issues. In contrast, the footpad health of Ross broilers was unaffected by SD. Other welfare indicators, including gait score, back and belly temperature, and bare-skin ratio, were not significantly influenced by SD in either strain. These findings suggest that welfare is generally well maintained within the SD-44 kg/m^2 recommended by the NCC [14].

4.4. Effects of Age and Sex on Broiler Welfare

Age significantly impacts all welfare indicators in broilers. As broilers mature, feather coverage increases, but feather cleanliness declines, and gait scores worsen due to increasing BW and pressure on joints. These trends align with typical broiler growth patterns and are consistent with the findings of Taira et al. [55]. The average footpad dermatitis score of Ross broilers was lower at 56 days than at 28 days, likely due to changes in litter conditions. At 28 days, wet litter adhered to the birds' feet, contributing to footpad dermatitis. By 56 days, litter naturally dried and caked litter fell off, allowing for the partial healing of footpad damage without manual intervention. Although litter moisture was not measured in this study, this observation underscores the importance of litter management in controlling footpad dermatitis. Similar findings by Taira et al. [55] suggest that transitioning broilers from wet to dry litter slows the progression of footpad issues. Future studies should include the continuous monitoring of litter moisture to confirm its relationship with footpad health and overall welfare. These findings reinforce the role of litter management in mitigating footpad issues and improving broiler welfare as birds age.

The impact of sex on broiler welfare indicators becomes increasingly significant as birds age, with females consistently showing advantages across multiple welfare parameters by 56 days of age. Females tend to have a lighter BW, which reduces the stress on their legs and likely contributes to improved gait scores and footpad condition. These differences may result from inherent physiological or behavioral variations between sexes, including disparities in BW, activity levels, and metabolism. Given that commercial broilers are typically raised as mixed-sex flocks, managing production separately for males and females may not be practical. However, understanding these sex-based differences can inform strategies to optimize overall flock welfare.

4.5. Limitations and Future Research Directions

The range of SDs tested may not fully capture the variety found in commercial settings, and trials with higher densities could offer additional insights. Further exploring practical applications such as tailored density strategies across different growth stages can provide producers with actionable insights. Developing adaptable management systems that adjust densities dynamically based on age or seasonal conditions could enhance production efficiency. The trials for the Ross 708 and Cobb 700 broilers were conducted months apart, which might have introduced seasonal variability, affecting the results. While some welfare indicators were measured objectively, others, like gait score and feather cleanliness, were manually assessed, introducing subjectivity. Additionally, litter quality, which significantly impacts welfare, was not systematically measured.

Future research should explore a broader range of SDs, both higher and lower, to identify welfare and performance thresholds. More precise measurements of litter conditions, including moisture and ammonia levels, could provide deeper insights. Cross-seasonal studies would help assess the interaction between SD and varying environmental conditions. Incorporating automated technologies such as image analysis for welfare assessment could reduce bias and improve accuracy.

5. Conclusions

This study comparatively investigated the effect of SD on the production performance and welfare of Ross 708 and Cobb 700 broilers. The results indicated that reducing SD improved feather cleanliness in Ross broilers and footpad health in Cobb broilers on day 56, although it led to a decrease in bone-breaking strength in male Ross birds. However, other key production parameters, including BW, FCR, gait score, body temperature, and bare-skin ratio, were unaffected by variations in SD. Overall, the current SD practices commonly used by farmers did not compromise the production performance or overall welfare of the Ross and Cobb broilers, suggesting that existing guidelines remain effective in balancing productivity and animal welfare.

Author Contributions: Conceptualization, Y.Z. and S.Z.; methodology, Y.Z. and S.Z.; software, S.Z., M.J., X.Z. and T.T. (Tanner Thornton); validation, Y.Z., H.G. and T.T. (Tom Tabler); formal analysis, Y.Z., X.Z. and S.Z.; investigation, Y.Z.; resources, Y.Z. and S.Z.; data curation, Y.Z. and S.Z.; writing—original draft preparation, S.Z.; writing—review and editing, Y.Z., H.G., T.T. (Tom Tabler), M.J., X.Z. and T.T. (Tanner Thornton); visualization, Y.Z. and S.Z.; supervision, Y.Z.; project administration, Y.Z.; funding acquisition, Y.Z. All authors have read and agreed to the published version of the manuscript.

Funding: This work is supported by the USDA-NIFA Agriculture and Food Research Initiative Inter-Disciplinary Engagement in Animal Systems (IDEAS) program (award No. 2022-68014-36663).

Data Availability Statement: The raw data supporting the conclusions of this article will be made available by the authors on request.

Acknowledgments: The authors highly appreciate the assistance provided by the staff and students at the University of Tennessee Knoxville (UTK), Department of Animal Science, UTK Johnson Research and Teaching Unit, and the USDA-ARS Poultry Research Unit.

Conflicts of Interest: The authors declare no conflicts of interest.

References

1. USDA. Livestock, Dairy, and Poultry Outlook: August 2022. Available online: https://www.ers.usda.gov/webdocs/outlooks/104508/ldp-m-338.pdf?v=5644 (accessed on 1 March 2024).
2. National Chicken Council. Chicken Industry Provides 1.5 Million Jobs and $417 Billion in Economic Impact: New Study. Available online: https://www.nationalchickencouncil.org/chicken-industry-provides-1-5-million-jobs-and-417-billion-in-economic-impact-new-study/ (accessed on 1 March 2024).
3. De Vries, M.; de Boer, I.J. Comparing environmental impacts for livestock products: A review of life cycle assessments. *Livest. Sci.* **2010**, *128*, 1–11. [CrossRef]
4. Scanes, C. The global importance of poultry. *Poult. Sci.* **2007**, *86*, 1057. [CrossRef] [PubMed]
5. Zuidhof, M.; Schneider, B.; Carney, V.; Korver, D.; Robinson, F. Growth, efficiency, and yield of commercial broilers from 1957, 1978, and 2005. *Poult. Sci.* **2014**, *93*, 2970–2982. [CrossRef] [PubMed]
6. Cornish, A.; Raubenheimer, D.; McGreevy, P. What we know about the public's level of concern for farm animal welfare in food production in developed countries. *Animals* **2016**, *6*, 74. [CrossRef]
7. Hall, C.; Sandilands, V. Public attitudes to the welfare of broiler chickens. *Anim. Welf.* **2007**, *16*, 499–512. [CrossRef]
8. Webster, J. Assessment of animal welfare: The five freedoms. In *Animal Welfare: A Cool Eye Towards Eden*; Blackwell Science: Oxford, UK, 1994.
9. Mellor, D.J. Moving beyond the "five freedoms" by updating the "five provisions" and introducing aligned "animal welfare aims". *Animals* **2016**, *6*, 59. [CrossRef]
10. Bessei, W. Welfare of broilers: A review. *Worlds Poult. Sci. J.* **2006**, *62*, 455–466. [CrossRef]
11. Tainika, B.; Şekeroğlu, A.; Akyol, A.; Waithaka Ng'ang'a, Z. Welfare issues in broiler chickens: Overview. *Worlds Poult. Sci. J.* **2023**, *79*, 285–329. [CrossRef]
12. Rayner, A.C.; Newberry, R.C.; Vas, J.; Mullan, S. Slow-growing broilers are healthier and express more behavioural indicators of positive welfare. *Sci. Rep.* **2020**, *10*, 15151. [CrossRef] [PubMed]
13. Meyer, M.M.; Johnson, A.K.; Bobeck, E.A. Development and validation of broiler welfare assessment methods for research and on-farm audits. *J. Appl. Anim. Welf. Sci.* **2020**, *23*, 433–446. [CrossRef]
14. National Chicken Council. National Chicken Council Broiler Welfare Guidelines and Audit Checklist. Available online: https://www.nationalchickencouncil.org/wp-content/uploads/2023/01/NCC-Broiler-Welfare-Guidelines_Final_Dec2022-1.pdf (accessed on 15 May 2023).

15. Mckenna, M. The Next Step in Animal Welfare? Breed a Better Chicken. Available online: https://www.nationalgeographic.com/culture/article/the-next-step-in-animal-welfare-breed-a-better-chicken (accessed on 20 April 2024).
16. National Chicken Council. Broiler Chicken Industry Key Facts 2021. Available online: https://www.nationalchickencouncil.org/about-the-industry/statistics/broiler-chicken-industry-key-facts/ (accessed on 15 May 2023).
17. Chan, I.; Franks, B.; Hayek, M.N. The 'sustainability gap' of US broiler chicken production: Trade-offs between welfare, land use and consumption. *R. Soc. Open Sci.* **2022**, *9*, 210478. [CrossRef] [PubMed]
18. Dozier, W., III; Gehring, C. Growth performance of Hubbard × Cobb 500 and Ross × Ross 708 male broilers fed diets varying in apparent metabolizable energy from 14 to 28 days of age. *J. Appl. Poult. Res.* **2014**, *23*, 494–500. [CrossRef]
19. Kryeziu, A.; Mestani, N.; Berisha, S.; Kamberi, M. The European performance indicators of broiler chickens as influenced by stocking density and sex. *Agron. Res.* **2018**, *16*, 483–491.
20. Vargas-Galicia, A.J.; Sosa-Montes, E.; Rodriguez-Ortega, L.T.; Pro-Martinez, A.; Ruiz-Feria, C.A.; Gonzalez-Ceron, F.; Gallegos-Sanchez, J.; Arreola-Enriquez, J.; Bautista-Ortega, J. Effect of litter material and stocking density on bone and tendon strength, and productive performance in broilers. *Can. J. Anim. Sci.* **2017**, *97*, 673–682.
21. Son, J.-H. The effect of stocking density on the behaviour and welfare indexes of broiler chickens. *J. Agric. Sci. Technol.* **2013**, *3*, 307.
22. Palizdar, M.; Daylami, M.; Pourelmi, M. Effects of high stocking density on growth performance, blood metabolites and immune response of broilers (ROSS 308). *J. Livest. Sci.* **2017**, *8*, 196–200.
23. Cengiz, Ö.; Köksal, B.H.; Tatlı, O.; Sevim, Ö.; Ahsan, U.; Üner, A.G.; Ulutaş, P.A.; Beyaz, D.; Büyükyörük, S.; Yakan, A. Effect of dietary probiotic and high stocking density on the performance, carcass yield, gut microflora, and stress indicators of broilers. *Poult. Sci.* **2015**, *94*, 2395–2403. [CrossRef]
24. Zuowei, S.; Yan, L.; Yuan, L.; Jiao, H.; Song, Z.; Guo, Y.; Lin, H. Stocking density affects the growth performance of broilers in a sex-dependent fashion. *Poult. Sci.* **2011**, *90*, 1406–1415. [CrossRef]
25. Yanai, T.; Abo-Samaha, M.; El-Kazaz, S.; Tohamy, H. Effect of stocking density on productive performance, behaviour, and histopathology of the lymphoid organs in broiler chickens. *Eur. Poult. Sci.* **2018**, *82*, 247.
26. Houshmand, M.; Azhar, K.; Zulkifli, I.; Bejo, M.; Kamyab, A. Effects of prebiotic, protein level, and stocking density on performance, immunity, and stress indicators of broilers. *Poult. Sci.* **2012**, *91*, 393–401. [CrossRef]
27. Gholami, M.; Chamani, M.; Seidavi, A.; Sadeghi, A.A.; Aminafschar, M. Effects of stocking density and environmental conditions on performance, immunity, carcase characteristics, blood constitutes, and economical parameters of cobb 500 strain broiler chickens. *Ital. J. Anim. Sci.* **2020**, *19*, 524–535. [CrossRef]
28. Aviagen. Ross 708 Broiler Management Handbook. Available online: https://aviagen.com/assets/Tech_Center/Ross_Broiler/Ross-BroilerHandbook2018-EN.pdf (accessed on 1 May 2022).
29. Cobb. Cobb Broiler Management Guide. Available online: https://cobbgenetics.com/assets/Cobb-Files/Broiler-Guide_English-2021-min.pdf (accessed on 15 March 2022).
30. Zhou, S.; Watcharaanantapong, P.; Yang, X.; Thornton, T.; Gan, H.; Tabler, T.; Prado, M.; Zhao, Y. Evaluating broiler welfare and behavior as affected by growth rate and stocking density. *Poult. Sci.* **2024**, *103*, 103459. [CrossRef] [PubMed]
31. Welfare Quality®. Assessment Protocol for Poultry. Available online: http://www.welfarequalitynetwork.net/media/1293/poultry-protocol-watermark-6-2-2020.pdf (accessed on 15 March 2022).
32. Sørensen, P.; Su, G.; Kestin, S. Effects of age and stocking density on leg weakness in broiler chickens. *Poult. Sci.* **2000**, *79*, 864–870. [CrossRef] [PubMed]
33. Zhao, Y.; Xin, H.; Dong, B. Use of infrared thermography to assess laying-hen feather coverage. *Poult. Sci.* **2013**, *92*, 295–302. [CrossRef]
34. Federation of Animal Science Societies. Guide for the Care and Use of Agricultural Animals in Research and Teaching. Available online: https://www.asas.org/docs/default-source/default-document-library/agguide_4th.pdf?sfvrsn=56b44ed1_2 (accessed on 18 October 2021).
35. Wang, B.; Min, Z.; Yuan, J.; Zhang, B.; Guo, Y. Effects of dietary tryptophan and stocking density on the performance, meat quality, and metabolic status of broilers. *J. Anim. Sci. Biotechnol.* **2014**, *5*, 44. [CrossRef]
36. Weimer, S.L.; Mauromoustakos, A.; Karcher, D.M.; Erasmus, M.A. Differences in performance, body conformation, and welfare of conventional and slow-growing broiler chickens raised at 2 stocking densities. *Poult. Sci.* **2020**, *99*, 4398–4407. [CrossRef]
37. Li, X.M.; Zhang, M.H.; Liu, S.M.; Feng, J.H.; Ma, D.D.; Liu, Q.X.; Zhou, Y.; Wang, X.J.; Xing, S. Effects of stocking density on growth performance, growth regulatory factors, and endocrine hormones in broilers under appropriate environments. *Poult. Sci.* **2019**, *98*, 6611–6617. [CrossRef]
38. Goo, D.; Kim, J.; Choi, H.; Park, G.; Han, G.; Kil, D.Y. Effect of stocking density and sex on growth performance, meat quality, and intestinal barrier function in broiler chickens. *Poult. Sci.* **2019**, *98*, 1153–1160. [CrossRef] [PubMed]
39. Abudabos, A.M.; Samara, E.M.; Hussein, E.O.; Al-Ghadi, M.A.Q.; Al-Atiyat, R.M. Impacts of stocking density on the performance and welfare of broiler chickens. *Ital. J. Anim. Sci.* **2013**, *12*, e11. [CrossRef]
40. Son, J.; Kim, H.-J.; Hong, E.-C.; Kang, H.-K. Effects of stocking density on growth performance, antioxidant status, and meat quality of finisher broiler chickens under high temperature. *Antioxidants* **2022**, *11*, 871. [CrossRef]
41. Ekstrand, C.; Carpenter, T. Temporal aspects of foot-pad dermatitis in Swedish broilers. *Acta Vet. Scand.* **1998**, *39*, 229. [CrossRef] [PubMed]

42. Hall, A.L. The effect of stocking density on the welfare and behaviour of broiler chickens reared commercially. *Anim. Welf.* **2001**, *10*, 23–40. [CrossRef]
43. Buijs, S.; Van Poucke, E.; Van Dongen, S.; Lens, L.; Baert, J.; Tuyttens, F.A.M. The influence of stocking density on broiler chicken bone quality and fluctuating asymmetry. *Poult. Sci.* **2012**, *91*, 1759–1767. [CrossRef] [PubMed]
44. Ma, H.H.; Xu, B.; Li, W.J.; Wei, F.X.; Kim, W.K.; Chen, C.X.; Sun, Q.Y.; Fu, C.; Wang, G.L.; Li, S.Y. Effects of alpha-lipoic acid on the behavior, serum indicators, and bone quality of broilers under stocking density stress. *Poult. Sci.* **2020**, *99*, 4653–4661. [CrossRef]
45. Bradshaw, R.H.; Kirkden, R.D.; Broom, D.M. A review of the aetiology and pathology of leg weakness in broilers in relation to welfare. *Avian Poult. Biol. Rev.* **2002**, *13*, 45–103. [CrossRef]
46. Tablante, N.; Estevez, I.; Russek-Cohen, E. Effect of perches and stocking density on tibial dyschondroplasia and bone mineralization as measured by bone ash in broiler chickens. *J. Appl. Poult. Res.* **2003**, *12*, 53–59. [CrossRef]
47. Buijs, S.; Keeling, L.; Rettenbacher, S.; Van Poucke, E.; Tuyttens, F.A. Stocking density effects on broiler welfare: Identifying sensitive ranges for different indicators. *Poult. Sci.* **2009**, *88*, 1536–1543. [CrossRef]
48. Rath, N.; Balog, J.; Huff, W.; Huff, G.; Kulkarni, G.; Tierce, J. Comparative differences in the composition and biomechanical properties of tibiae of seven-and seventy-two-week-old male and female broiler breeder chickens. *Poult. Sci.* **1999**, *78*, 1232–1239. [CrossRef]
49. Applegate, T.; Lilburn, M. Growth of the femur and tibia of a commercial broiler line. *Poult. Sci.* **2002**, *81*, 1289–1294. [CrossRef]
50. Alkhtib, A.; Sanni, C.; Burton, E.; Scholey, D. What is "normal"? Morphology and mineralization of tibias from healthy, on farm broilers. *J. Appl. Poult. Res.* **2021**, *30*, 100190. [CrossRef]
51. Mönch, J.; Rauch, E.; Hartmannsgruber, S.; Erhard, M.; Wolff, I.; Schmidt, P.; Schug, A.R.; Louton, H. The welfare impacts of mechanical and manual broiler catching and of circumstances at loading under field conditions. *Poult. Sci.* **2020**, *99*, 5233–5251. [CrossRef] [PubMed]
52. Lingham, S.T. The feather, a triumph of natural engineering and multifunctionality. In *Poultry Feathers and Skin: The Poultry Integument in Health and Welfare*; CABI: Wallingford, UK, 2019; pp. 3–11.
53. Stamp Dawkins, M.; Donnelly, C.A.; Jones, T.A. Chicken welfare is influenced more by housing conditions than by stocking density. *Nature* **2004**, *427*, 342–344. [CrossRef] [PubMed]
54. Meluzzi, A.; Fabbri, C.; Folegatti, E.; Sirri, F. Survey of chicken rearing conditions in Italy: Effects of litter quality and stocking density on productivity, foot dermatitis and carcase injuries. *Br. Poult. Sci.* **2008**, *49*, 257–264. [CrossRef] [PubMed]
55. Taira, K.; Nagai, T.; Obi, T.; Takase, K. Effect of litter moisture on the development of footpad dermatitis in broiler chickens. *J. Vet. Med. Sci.* **2014**, *76*, 583–586. [CrossRef]

MDPI AG
Grosspeteranlage 5
4052 Basel
Switzerland
Tel.: +41 61 683 77 34

AgriEngineering Editorial Office
E-mail: agriengineering@mdpi.com
www.mdpi.com/journal/agriengineering

www.ingramcontent.com/pod-product-compliance
Lightning Source LLC
Chambersburg PA
CBHW062044220926
51850CB00010B/2655
* 9 7 8 3 7 2 5 8 7 6 3 3 4 *